Praise for *Material*

'In *Material*, Nick Kary mines the deep knowledge of makers and creatives, and the resultant nuggets from carpenters, weavers, smiths and masons are often gold. This book offers a timely retort to a world in thrall to fast, ephemeral fashions. A marvellous mix of the heuristic, didactic and expedient; a celebration of the local, old-school and hands-on, *Material* is generous, wise, fascinating and fundamentally humane.'

—**Dan Richards**, author of *Outpost*

'A profound and personal delving into the ancient connection between man and matter and our increasingly tenuous relationship with nature. An important book, brimming with insight.'

—**Nicholas Evans**, author of *The Horse Whisperer*

'Nick Kary understands that to be a maker is to be a seeker – creating to a personal standard not only out of the material of the earth but of memory, one's relationship to place and history, the force of time. The work of our hands affirms a stewardship of the land that is also an imperative. *Material* is a quiet, heartfelt assertion of why craft so deeply matters.'

—**Anne Michaels**, author of *Fugitive Pieces*

'It was as if Nick Kary's outstretched hand took mine and, tucking my arm under his, gently led me into an enchanted world. There is an exquisite poignancy in this book, an honesty, a fearless enquiry that shifts from sunlight to shadow, along paths mostly hidden from a world grown weary of beauty. Material meets maker in a sensuous weave of insight, wonderment, ordinariness, and deep humanity. I will read this book again, and slowly, in the way I might cup hands and, dipping them to clean spring water, pause to drink.'

—**Mac Macartney**, author of
The Children's Fire: Heart Song of a People

'With grace and humility, Nick Kary has crafted a deeply felt and intimately observed portrait of a magic English landscape of authentic makers working amidst the remnants, scars and generational stories of forgotten crafts and industry. *Material: Making and the Art of Transformation* beautifully weaves together the pathos and promise of traditional materials and methods, and the intimate bonds that form between artisans and their medium as they bring meaning to their making. Nick Kary has gifted us, giving eloquent voice to thinking and feeling with one's hands.'

—**Christopher Bardt**,
author of *Material and Mind*

Material

Material

Making and the Art of Transformation

Nick Kary

Chelsea Green Publishing
White River Junction, Vermont
London, UK

Project Manager: Alexander Bullett
Commissioning Editor: Jonathan Rae
Developmental Editor: Benjamin Watson
Copy Editor: Eliani Torres
Proofreader: Susan Pegg
Designer: Melissa Jacobson
Page Layout: Abrah Griggs

Printed in the United States of America.
First printing September 2020.
10 9 8 7 6 5 4 3 2 1   20 21 22 23 24

Our Commitment to Green Publishing
Chelsea Green sees publishing as a tool for cultural change and ecological stewardship. We strive to align
our book manufacturing practices with our editorial mission and to reduce the impact of our business
enterprise in the environment. We print our books and catalogs on chlorine-free recycled paper, using
vegetable-based inks whenever possible. This book may cost slightly more because it was printed on paper
that contains recycled fiber, and we hope you'll agree that it's worth it. *Material* was printed on paper
supplied by Sheridan that is made of recycled materials and other controlled sources.

Library of Congress Cataloging-in-Publication Data
Names: Kary, Nick, author.
Title: Material : making and the art of transformation / Nick Kary.
Description: White River Junction, Vermont : Chelsea Green Publishing, 2020.
Identifiers: LCCN 2020026889 (print) | LCCN 2020026890 (ebook) | ISBN 9781603589321
    (hardcover) | ISBN 9781603589338 (ebook)
Subjects: LCSH: Handicraft—Philosophy.
Classification: LCC TT149 .K37 2020 (print) | LCC TT149 (ebook) | DDC 745.5—dc23
LC record available at https://lccn.loc.gov/2020026889
LC ebook record available at https://lccn.loc.gov/2020026890

Chelsea Green Publishing
85 North Main Street, Suite 120
White River Junction, VT 05001

Somerset House
London, UK

www.chelseagreen.com

But who dreams the bird in
the wild wood

with the berry in its beak that
becomes the tree

that the woodcutter fells his
eyes lit with the form

his hands seek in the wild grain which he
hews and shapes

into his chair, poised now in
the winter gleam

of a quiet room?

Chris Waters, 'Chairmaking in Yew'

Contents

Acknowledgements

This book owes its life to the belief of others. To my editor Jon Rae for his unflinching support as the book slowly strayed ever further from its initial description. To my daughter Misha for her daily questions over the last year as to whether my writing has gone well. To my wife Dolly for believing in me, for her direct critical feedback, her reading and rereading, and her absolute commitment to the project – and tolerance of my moods. To Tara and the staff at the Green Table café for supplying me with food, coffee and humour. To my writing friends Chris and Pip for their constant feedback and support, to my friend Michael for always being there and to Mac for reminding me to trust myself. To Jay Griffiths for her honesty and words of wisdom. To Ben Watson, Eliani Torres and all the staff at Chelsea Green who have supported this book from its inception to its completion. To my sons Felix and Oscar for their commitment to their own visions, and to all my friends and family for their support.

The writing of this book would not have been possible without the generosity of all the makers who have allowed me to spend time with them in their private spaces, to share with me their thoughts and their reflections. I particularly thank Lin Lovekin, as her generosity and the invitation into her home helped open my vision for what this book could become. I thank Bernard Graves of Pyrites for letting me join his craft camp for the week, all the facilitators who made time to speak with me, and Rich and Mark for accepting my presence in their course. Finally, I thank Lou Tonkin for her beautiful illustrations and for understanding so intuitively what I was looking for.

Further to these individuals are all the other makers, dead and alive, whose stories feature in this book: Charles Bedford, builder, carpenter, sniper and grandfather; David and Paul Lethbridge of Barton Sawmill; Mark Wills of Blue Hills Tin; Hilary Burns, basketmaker; Geraldine Jones, basketmaker; Mike Gardner, woodsman, timber framer and sawmiller; Naomi the coppicer; Henry Adams of Buckler's Hard, shipbuilder; Nic Collins, potter; Sabine Nemet, potter; Jonathan Garratt, potter; Richenda Macgregor, potter; Lucy the apprentice potter.

Dave Budd, toolmaker and smith; Walter Kary, photographer, weaver, liaison officer and grandfather; Greg Rowland of Mike Rowland and Son Wheelwrights; Steve Overthrow, riddle maker; Philip Kazan, novelist and cook; Claire Williamson, poet and educator; Chris Waters, poet and woodworker.

Introduction

A little while ago I had an idea for a bed, sketched it down roughly and started making the pieces for it. I'd selected some Sycamore a friend of mine gave me from his woods, creamy white and beautiful to work with. I enjoyed the work, making each piece using only hand tools and traditional methods. I loved the dance of the making, of revealing the beauty of the wood while not imposing my will on it too forcibly. Gradually, the pieces came together into various sections of the bed. It looked like my sketch, like the image in my head, yet for some reason it failed to satisfy me. The bed remains incomplete in my workshop, an adornment to the space, a reminder that not all ideas achieve what we would hope of them.

This book is a little like that bed; the earlier sections of it now discarded in a file on my laptop, the hopes I'd had for them and their place in these pages no longer relevant. This book did not become what I expected. It may even have become something I thought I wanted to avoid. Yet in all truth I'm not sure I ever really knew what it would be like, holding to an idea only because it seemed true to an intention.

It is an odd thing for a pragmatist like me to say, but the call I had to write this came from a place deeper than my consciousness could understand. As an experienced craftsman, I thought I would probably be writing a book about making, so I was surprised by the title that kept forcing its way into my mind. It took me a long time to understand why I wanted to call the book *Material.* When the title first came to me, early in the project, I tried to fight it, and as the writing progressed, I kept expecting the name to change; but it refused to do so.

I had always taken the material of my making for granted, relating to it in the plural, a choice of inanimate 'materials' that were at my disposal. Yet now, the word presented itself to me in the singular, with all the gravitas of something much greater than racks of planks, metal rods or pieces of leather awaiting transformation. It asked me not only what *a* material was but also what was *material;* it forced me to look at my work, and that of all of us who enter into a relationship with the materials we use to make things. It forced me to look at where these materials come from, at the often untold stories of their extraction from the natural world, and at the scars and consequences they leave behind.

The first thing I had to do once the project became clearer to me was to accept the personal nature of it. The material I work with is *material to me,* built on my personal relationship with landscape and transformation, and the dignity I wish for myself and the ground I inhabit. As a maker of wooden objects, I cannot get away from the 'nature' of the material I work with – that is, the nature contained within it. I write this book from the perspective of a maker but also as a human being, part of the species that has collectively wrought the greatest damage on this planet. The maker in me is inseparable from the human, for it is what has distinguished us from other species. Our capacity to make – to alter and transform the materials we extract from the natural world – is our gift and our curse. I feel both aspects intensely, and this book is my attempt to look more deeply into that duality and make some peace with it.

Making is still the core of the book, and the makers I have hung out with while writing it have continually been my guides back to the material, to the earth and to what underlies the making process. My journeys down underground, both physical and metaphorical, through mineshafts and history, started in conversation with those who make objects or process materials.

I imagined that I would explore a wide geographic area from which I would harvest stories from various makers, but instead I have been guided more by the landscape of the region where I have lived for the last twenty years, South West England. Writing this book taught me just how much I have taken for granted, how ignorant I have been about what lies right around me and how uneducated I really am. Quite a realisation from someone who apparently had the best education one can receive and who may be perceived by others to be 'well educated'. The writing and research actually made me feel quite ignorant and narrow, blind to the truths bleeding from the wounds under my feet.

The roots underlying the narrative structure of the book grow from the ground of my own relationship with making, and where I live and what matters to me. These roots bind me to a sense of place, but from here they spread out amongst the roots of others, through the mycelia and hyphae, interconnecting with the stories of other makers. They tap through the water and rock under our feet and find their way through Cornwall, Devon, Dorset and Hampshire. They explore the movement of landscape, the granite rising from the Earth's core and the legacy it has left to this part of the world. The minerals caught within it, the formations left behind by erosion, and the woodlands scattered around its margins have been my guides, alongside the makers working there.

I tried to invoke some sense of order, but order has a habit of continually being re-formed by the chaotic and creative moment. The book is a journey from my own personal relationship with making to those of others via the landscape around us. I have not attempted to create a systematic narrative structure or be definitive in any way. Rather, the book is composed of various stories, which I hope will knit together for the reader in a narrative that allows for their own stories and experiences.

My father followed in the footsteps of his father, and grandfather, in the running of a cloth mill on the outskirts of Vienna. He could not resist the pull of his former vocation and brought me up to discern the quality of a piece of cloth by feel. He taught me to rub the fabric between thumb and forefinger and sense the soapy softness of texture he valued, and which linked him back to the thread of connection that he had lost. I thank him for this gift; it has served me well to discern quality with touch and not by sight alone.

I, too, am pulling at a thread, following it through and seeing to what it is attached. I am attempting to find the warp under the fabric of our making relationship with the natural world, and to see and understand the colours and patterns of the yarn that weave across it. These patterns give the illusion of an object: carpet, rug or basket. In all truth, those threads – dyed naturally or artificially, of cotton or of Willow – are simply a reduction of natural resources, and the object they create is simply an ordered reassembly of those resources in a form that suits our needs and desires. The making of this book, the making of a piece of furniture, and the evolution and structure imposed on our world by humankind, are all parallel pathways in this exploration. They are the warp across which I seek to weave coloured threads of diverse topics that are inextricably bound to the cloth.

My enquiry into the processes of making and material use is inextricably bound to less tangible questions about time, belonging, beauty, perfection and the past. The thread I pull at seems so often to be the same one. What is it that we are part of, both in relation to where we have come from and where we find ourselves? By what means, I ask myself, are we made? Family, its cultural perspective and societal group, is an inescapable base from which my ideas have been shaped. This library of codes, conventions and narratives establishes who we are and the choices we may be led to make. Wherever our lives take us, our origins remain in the background, a memory reserve and an experiential blueprint. Yet our choices in any given moment help to define a narrative uniquely our own, and for me personally this has been guided by finding the maker within.

Chapter One

Roots

We talk of things having 'material significance' or 'material substance' as a way of indicating the importance we attach to them. We talk of matter, and what matters, what the earth is composed of and what is of relevance to us. We talk of materialism, often in relation to consumption, perhaps forgetting just how linked to material our lives are by necessity. The word has come to us through the Latin *materia*, which itself derives from the Latin *mater*, meaning 'mother' or 'origin'. The meaning of the feminine noun *materia* ranges from 'breeding stock' to 'matter', and includes 'fuel', 'wood', 'material', 'latent ability' and 'potential'. So, from mother we have the idea of origin, where we are from. It speaks of the latent ability or potential of a person or thing as if they matter.

I write this book to help me understand the relationship that we all have fashioning our lives within the material structure of this planet. Looking back to see how a word has changed meanings over time, what its roots are, helps me in this process. Material is understood today to represent a physical form, something not 'spiritual, mental or supernatural', but rather 'real, ordinary, earthly, drawn from the material world'. As such it is seen as separate to the human, something which does not have an effect on us, but something we can have power and control over. This is quite different to its origin from mother.

What matters to an individual or community helps ground us personally and collectively and becomes imbedded in the varied cultural narratives of our society. Within and around us, matter forms the fabric of all we see, the materials to which we relate throughout our lives. An object made of physical matter and particles may have material significance to a person because of what that object represents to them in the form of personal history. I see our belonging here, as the roots which bind a tree to the Earth. My sense of my material existence is through the roots that bind me to the life I live.

There was a moment many years ago when I took a leap of faith and sought to find a connection to something that 'mattered'. I had just finished my degree and wondered what it had all been about. The only solid, powerful thing I could connect with was the ground beneath my feet.

In that moment I remembered my English grandad Charles Bedford and his earthiness, and how much I enjoyed the tinkering we had done together. I knew I needed to become a maker. I needed a physical connection to the material, and not more mind-centred meandering. I needed to touch and feel the body of a thing, and I wanted to learn how to transform wood into objects of beauty and utility. I needed centring, and the dignity and physicality of the making process seemed to be my answer.

Roots

I embarked on a road whose curves and undulations were unpredictable. I learnt the hard way, first as a designer, manufacturer and supplier, and later as a designer-maker. This is the path I have remained on, originally trained through apprenticeship and at college, and then by the lessons of what works, and what doesn't. My body has slowly learnt a new language, and my mind has grown beyond my brain. I now make furniture crafted from local hardwoods, each piece bearing the marks and character of the original tree. When I was trained we removed all the 'imperfections' that were left in the often imported woods supplied to us straight-edged and uniform. The idea of what was perfect seemed almost unattainable, far away from the dictionary definition of 'bringing something to completion'. The years of relationship with my craft have been a journey of discovery, and the idea of what should be has slowly dissolved into the simplicity of what is.

I think often of my English grandad, as it was with him as a child that I began to experience the joys of tinkering in the workshop. He died when I was eighteen, and in the normal self-obsessed behaviour of a young man, I did not dwell on him much at the time, but a few years later, just before I graduated, the memories of him flooded back.

I grew up in a time when feelings were rarely expressed in words, when love, pain and difficulty were often communicated through behaviour. My most enduring recollection of the love my grandfather expressed for me was of moments when I would have been somewhere between six and fourteen. My mind has compressed these moments into a single memory:

Material

We are having afternoon tea in the living room of my grandparents' small, semi-detached house in Hertfordshire. My grandmother Maud Bedford, in her flowered apron, is putting a plate of freshly baked biscuits on the small mirror-polished Mahogany side table. Tea is served from the china teapot in little porcelain cups that I could never quite work out how to hold. My grandad's nicotine-stained fingers stand out in my memory as he quietly sits there, cup in hand, silent and thoughtful. My grandmother fusses and chats, the sweet smell of her baking still wafting in from the open kitchen door. All the while I am wondering if my grandad has forgotten his promise, whether his withdrawn state will give way to sleep, leaving me alone amongst the polished brass and darkened beams.

Yet with perfect grace and timing he excuses us both and encourages me out of the living room. I follow this kindly and enigmatic man, excited by the prospect of what lies ahead. We leave the kitchen, my grandma's domain, replenished by her material display of love. The biscuits were always a sweetener, a manifestation of what couldn't be put into words.

Soon we are outside, the well-nurtured rear garden and hidden pond a testament to the old man's love of making. We slip through the side door of a 1930s garage with a narrow workbench on one side, just enough room to accommodate his old grey Ford Anglia. That very car would put an end to his driving ambitions, as some years later he would drive it right through the rear wall of this garage, after which it arrived at our house to become the tortured object of my own driving ambitions. It isn't in his garage now, just the long, well-used bench, two vices attached to it.

A smell I am so familiar with, yet have no adequate words for, hangs in the air, something between used car oil and fresh wood dust. My grandad is standing by the bench, his movements slow and deliberate, his fingers busy in his tobacco tin, rolling another cigarette (one of sixty or more a day, a habit that ultimately led to his death). It is an intent, meditative action, assured and dextrous, honed by constant repetition.

Soon the tobacco smoke is in my nostrils, filling the garage, and to this day it gives me comfort. It represents measured behaviour and slowing down and calming. Before I start working in my shop, I will often smoke a pipe in his memory, for the feel of the bowl, the connection to the process and the invitation to calm. I feel the warm bowl in my hand as I write this, the smoothness of the burl, the tricks of light its grain plays on my eyes, its material presence helping connect me to mine in this moment.

Only now does he allow me to see a wicked grin. From below the bench he deliberately pulls out another green tobacco tin, this one clanking sharply as he places it on the oily benchtop. Opening it, he turns to look at me, the half-smile back again, both of us complicit in what is to come. The door is shut and we are alone. I know that what is to come is a performance. It is for me. Yet it is of and from him, as it will speak of his past for which he has no words.

The lid is off, lying on its back alongside the open tin. Inside the tin is a small selection of pointed brass tubes, bullets that he kept from the First World War, in which he had served. He served at Gallipoli and in Africa as a sniper in the First War, and in the Second World War he was part

11

of the Home Guard and a stalwart member of the local community, training volunteers in the safe demolition of damaged houses. He was awarded the Military Cross in the First World War, and commended elsewhere for his marksmanship and carpentry skills. He never accepted the award as he disapproved of the inequality between the rations given to officers and soldiers.

He carefully takes one bullet and places it in the jaws of the metalworking vice. The pointed and deadly end stands up above the jaws. Aware of my absolute focus on his movements, he takes delight in his performance, slowly reaching for a pair of burnished steel pliers that were laid neatly on a shelf above the bench. He takes them, places them around the head of the bullet and in one deft movement removes it from its casing.

My heart is pounding, for even though I have witnessed this trick before, its sheer bravado is terrifying to me. Recovering myself, I look to find him offering me a small ball-peen hammer, its flat side facing down as he places it carefully in my hand. He then releases the casing from the vice, turns it upside down and retightens the vice. With an intense look in my direction, he picks up a centre punch and places it on the firing cap, nodding to me that it is now my turn. 'Come on, lad,' his eyes say in their kindly way.

I raise the hammer, nervous that I should hit it square, and, hovering momentarily, I bring it down on the punch, striking it a little askew. There is a sharp uttering and a rush of air as the strike ignites the explosive, and a small puff of smoke wafts away from the vice. I look at him with some alarm, and he, catching my eye, mischievously lets out a restrained yet infectious peal of laughter.

When I recall this moment and the many others like it, I am caught somewhere between nervous tension and ecstatic delight. Soon we are at the threshold of the kitchen again, welcomed by the stern look of my grandma. She scans us from head to toe on our re-entry, to ensure that we will both thoroughly clean ourselves. That memory has never left me, and by retelling it to my children, I make sure that it never will.

I feel connected to my grandad through this memory, but also through the object, the bullet. The bullet's matter has become the matter of my memory, and with it my memory jumps across the English Channel to the land of my other grandfather Walter Kary. While one grandfather fought the Turks at Gallipoli, the other fought the Russians in the Carpathian Mountains and Eastern Poland. I never knew him, nor did my father. Yet I do have in my possession six hundred glass stereoscopic slides taken on the Russian front while he served as an officer in 1914–15. These slides are material for my own constructed narrative about him, and, along with some domestic pictures, they are all I have to remember something for which there are no known stories.

I take a similar view regarding what I have lost in terms of relationship to the natural world around me and my interaction with it. As a young man I had a feeling of disconnection, of a profound separation between the self I identified with and a sense within me that there was something more, something greater for me. This sense led me to become a maker, and has thus connected me to my body and spirit and to the materials I use, where they originate from, and the impact their extraction may have. So, as I sought to reconstruct a relationship with my fractured social history and family narrative, I have also sought to reconstruct my relationship with the physical world around me. The question that repeats itself is: What is the material I work with, and how does my engagement with it affect the material of my own existence, my own place here?

Thinking of my grandfathers now, the one born in 1896 in England and the other in 1876 in Vienna, I feel connected to a time when making happened within communities, when villages each had a farrier, a carpenter, a saddle maker, a butcher and a baker. Materials sourced locally were made into objects by local craftsmen. Women knitted sweaters, made clothes, bottled and jarred the harvest, and children witnessed it all, already apprentices to their material relationship with life.

In his memoir *The Village Carpenter*, Walter Rose, writing in the 1930s, speaks of the era before the First World War and about his father's carpentry shop in southern England. He writes knowing he must document his memories before he dies so that there will remain a testimony of that time. He speaks of the various jobs these carpenters did, from the making of windows and doors, the framing of houses, the building of roofs, staircases and furniture, to the making of coffins, carts, cartwheels and even the Elm pipes for wells. He speaks of the relationship individual craftsmen within the shop had to particular aspects of the work, and how that connected them to the community.

The two men who went out to the farmers' fields to repair the Chestnut fences and gates, the men who had a particular eye for the precise task of drilling a straight hole 12 feet through the trunk of an Elm tree to make a well pipe. Then there were the sawyers, often itinerant men who would cut through the great logs. The top sawyer guiding the blade, the bottom one, not known for his subtlety or acumen, who pulled the saw blade down with force, its teeth cutting through the wood, the wet sawdust showering his face and body.

George Sturt, writing *The Wheelwright's Shop* some years later, further imbeds this picture of interconnected relationships. The early, dark starts, the wind howling out of the dawn into the cracks between windows and doors, the wood drying ready for use, the sense the craftsmen had of the purposing of each piece. The trees, sawn and stacked, became cradles for birth and coffins for death, and, between the two, cartwheels turned, taking all the industry of the carpenters' shops from the tree to its end purpose within the community.

Slowly but surely, metal pipes replaced the Elm ones, barbed wire the traditional fencing, and sawmills were established outside the carpenter's shop. Coffins came to be made of man-made materials instead of Elm, and gradually the quality and extent of the village carpenter's craft diminished. Eventually, the village carpenter also disappeared, to be replaced by large centralised workshops and factories in towns and near major roads. Alongside his disappearance was that of the village smith, wheelwright, saddle maker, butcher and baker as communities became fractured and a new order took hold. The car replaced the cart, the engine replaced the horse, and metal replaced wood. Objects and dwellings were no longer crafted by local people from local materials. The direct connection between place and material object was gradually fractured.

———

I have come to think through making, am inextricably bound to it, but also to all the contradictions implicit in that relationship. Now there is a growing thirst and interest in what it means to make, and I can't help but reflect on the longing I had thirty-five years ago and on what I see and hear around me now. Having made the decision to alter my path and ground myself in the journey to become a maker, I never imagined I would spend thirty-five years in one vocation.

It is tempting to ask why I have made a life of something so physical and demanding. Only recently, particularly through the traumatic but cleansing event of my workshop burning down, have I come to find an answer to that question. I have developed the ability to transform material into any form I wish within nature's laws. I have built our own home and filled the homes of countless others with furniture. I and all those who make are to some degree magicians, alchemists, the transformers of base nature into something undeniably human. This is not something I have understood before. I have simply made to live, unaware of how essential making is to the idea of being human.

As a maker I feel my humanity, my human *being*, the potential and danger I represent. The home that we have built for our family is a particular result of this dual relationship. We live in a woods where we have built our house and cabins, our studio and workshops. The woods up here hold the water, act as conduits for it and cling in their deep brackish mulch to it over summer and winter. It rises wilfully, raw freedom from the earth, and all becomes soft, musty, musky, fetid with deep life. The Sycamores, Beeches, Chestnuts and Oaks cast their seeds in its receptive womb, and the unstoppable, unnameable life force rises anew from it each season. We, on our small plot, sit in the middle of 20 or so acres of mainly Beech woods, planted a hundred years ago on a ridge, to protect the crops from sea- and moor-birthed gales. A hundred years of their growth, and whatever preceded them, have cast the ground brackish; our property, The Brake, taking its name from the soil.

When planted, this part of the ridge was in the middle of a 2,000-acre estate, two grand houses belonging to the family on either side at some distance from what is now our house, which was built in 1909, the same time as the period that George Sturt and Walter Rose are looking back at in the writing of their books. It was a timber bungalow, shipped flat-pack from the USA, erected here amongst the 20 acres of newly planted Beech trees, for the son of the estate and his young family. I have a magazine photo showing identical buildings, advertised to the British market to be shipped to the UK. Placed lightly on the ground, brick- and slate-propped, it was gradually enclosed by growth and the damp, shaded inclination of the land. The son was killed in the First World War, and the house witnessed the mass evacuation of all the land to the south of it as the Americans came to practise for the D-Day landings in 1944. By then it had been cut away from its land, a new house built next door, sold away from the family, who had retained only the 5-acre woods. So it remained for years, encapsulated by the growing trees, ringed and shaded in entirety by the time we moved here, invisible to all except those who ventured deep within.

Freshly erected after its journey from the USA by ship and horse-drawn cart, our house was called Ranch Bungalow, a wooden house sitting on a great plot overlooking the moors to the north and the sea to the south. Years and years later, the land round it was sold off, subdivided and sold again. Renamed The Brake for the ground on which it stood, the little house holds a very different story. It was cut off from its past, from the time that birthed it to the time we came to be its caretakers. From 20 acres to two-thirds of an acre; from the Ranch Bungalow, dominant over its land, to The Brake, enclosed by trees and the wild will of nature's growth.

How does a house fit on the back of a horse and cart I wonder, all the frames small enough and big enough, a detailed plan somewhere for how it would all fit together? The plot, flat as it was, would have been dug over by hand, local people on day rates digging trenches, a bricklayer laying a narrow footing around the perimeter, placing slate on top as a damp-proof course, much needed given the wetness of the land. In 1909 British tree stocks were already on their knees, to suffer even more by the end of the First World War. With little supply of native conifers for good building wood, we were already relying on imports from Scandinavia and the USA. The idea of sending a house flat-pack from so far away across the Atlantic seems nuts, yet at the time it may have been sensible, the timber and skills so much more innate in the USA than on home soil.

To build a house from the land, a house of cob – clay dug from the ground, rubble stone from local farmers' quarries – would have been the way here once. It would have taken the community to build such a structure, the sheer weight of clay and straw and dung mixed, the stone and lime, the mass of it heavy on the wet land. That was the tradition here, in the clay-lined Devon hills and valleys, with little woodland and a limited brick industry. It had been the way when the ancient farm houses were built in the lee of hills, on the edges of the valleys. It had been the way for the mill-houses and workers' cottages. These buildings had risen out of the earth on which they stood, with little in the way of damp-proof course, only the bottom layer of rubble stone to prevent moisture from wicking up into the cob walls. Damp outside, and damp inside, part of the wild, and it finding its windy way into the sanctum of the home.

Our house, together with its American name, was a house of the future, a house of foreign birth, of American efficiency, of the Raj whence the term 'bungalow' travelled to Britain. It was in its name and origin a cry to the future and a cry to the past, to the dying days of the British Empire and a shift in the power structure of world trade. The timber it was built from came from far distant trees, cut on great mills, far from these valleys, far from our consciousness. The techniques used to frame it, repetitive slight wooden uprights supporting lightweight horizontals, the whole becoming strong only when assembled, was far from the clay and rubble stone solidity of the Devon mind.

I think of the spades and shovels of the men digging the shallow trenches of the light wood structure in 1909. I imagine they were a little incredulous for the ease of it all, for the sheer simplicity of what they were providing for the house to languish on. If the house were plucked up from its little brick surround today, flown away from here, it would leave little sign of itself that a day of work would not remove. If left for a long time, if untended and unloved, all our wooden structures here would gradually become reabsorbed by the damp fertile woods. The leaves would pile uncleared at their feet, the moisture wick itself up into their frames, and with time the growing mulch would creep up, reintegrating the wood into the earth, taking away the structure, and slowly allowing the roofs to collapse back onto the ground. There they would sit, slower to decompose, absorbed by the green growth that would find the earth amongst them. They might become like woods I have seen, a mausoleum of the human condition, as trees grow around fridges and old cars, bicycle wheels emerging out of their trunks.

This is home. To make our home on land we have the privilege to be able to build and live on. To build home, and nurture home, is to maintain our human need amidst nature's will to repossess. To make home is to take land from the wild green will that licks around the edges of all our lives. For us to make home can mean that we deprive other creatures of theirs. Home is material, of material and material to our well-being. Home for me is safety, family and belonging. Home is a longing that pulls me to it, that asks of me how do I wish to live my life. Home is a privilege, as is safety in this wild world of ours, to be safe amongst the winds, amidst the arid breath of the hot regions or the cold air of northern climes. To be safe and held while we hold the balance of the natural world around us. I hope that if we stamp our feet into the earth, the print we leave is as soft as it can be, that the material we stamp our image onto can find its way back to the life force that created it.

Our word for home is derived from the German *Heimat*, which means not only 'home' but also 'the place where one is from'; the word connotes belonging and provenance. I had never lived anywhere for more than five years before I came here, and over a period of twenty years there is ever more a feeling of belonging. This home we have, this plot of land that we have possessed, filled with the material of our endeavours, finds itself on a ridge of sedimentary rock formed hundreds of millions of years ago. It rises up above the River Dart at its mouth, and runs north-west, curving slightly back south towards Kingsbridge, losing momentum, the weight of time pressing at it. In my ignorance I would imagine water running down from a high ridge, down the gouged and hidden Devon valleys that fissure this land, down into the River Gara or the River Avon, to the River that runs to the sea.

Yet it is not so, or at least not only, for this ridge seems to draw the water out of its rock core, capillary forces sucking it up and belching it out into the clay blanket that shrouds us. The water draws up when the weather is wet, the saturated ground bubbling up occasionally. The builders of the house dug down a well into the great spring of the land, tapped like a root into the moisture to draw out the water for their needs. The well rises and falls through the year, never empty, much of the time only inches from the edge of the earth line.

The drive from the main road cuts steeply up through the Beech woods, the trees' boughs crossing high above it, the foliage thickening the quiet of the air, softening the road's drone. Twenty years ago we drove up this drive for the first time, up into the lives of the family living here, our own dreams packed into our hopes, a rattling promise of possibility. Leaving the car on the dirt of the yard, we were both silent in the woods, awed by this place, surprised; for all our expectations, we could not have pictured the aura, the feel of the ridge, the timeless, still quiet of an unsung hymn.

Wild roses, pink in the light filtering through the conifer to which they clung, marked our entrance through the trees' shade. Cut green grass in our nostrils, the textures of vegetative life abundant in the southerly light of the unfettered sun. The house might not have stood there at all; it was as a shadow clinging onto the edges of a promise, an interloper amidst the wild abundant growth. Massive Rhododendrons had marched through the garden unchecked, encroaching from east and west, their violet pink blooms showing themselves off for our visit. A glimpse through them, across the grass and scattered shrubs, made faint promises for a world outside this secret garden on its island ridge.

The Beech trees to the west rose as a wall off the boundary edge only 20 metres from the house. Their years of growth towered in as many feet as they were old, 1909 to 1999, topped with crowns dense and abundant to mark their dignity. After midday, light could barely find its way through their leaves; little apart from the heavily clustered Rhododendrons could grow in their shade. The must of the mulch rising from the ground, the years of its transformation from fallen leaves to earth, hung thick in the air. In the house the smell of rot was more acrid, barely disguised by branded sprays, thick and clagging to the breath. The house was dying, on its knees with the stench of death. The health of it had affected those living there, drawn at their breath, robbed them of it as oxygen fed the rot, the wet earth seeking to repossess.

Our friends thought us mad. We bought the house for its promise, but it was a personal promise made in silence; no one else aside from us witness to it. Those friends helped us drag the carpets from it, exposing the rot that the five layers had held together, driving us gasping from the house. My wife could not breathe and choked asthmatically on the fungal spores that rose thickly into her lungs. We could not live there, and so we built a cabin not far from the house – dry, rot free and our home for the next six months. Simon, Matt and Russ dragged the entrails from the bungalow, stripped it of its floors, ceilings, substructure and internal walls. They tore out the lean-to where the rats had lived more readily than the humans; they tore out hundreds of feet of rotten wood, more like sponge than the stuff of trees, filled with disease, wet or dry with the oxygen that had been robbed from the underbelly of the structure.

The house had been added to over the years, added to in ignorance of the living being that it was. The fungi that would be breaking down the densely packed fallen leaves to mulch arrived under air-starved floors; the damp had nowhere to go, no air swept through by a cleansing breeze. In this house the fungi had bred happily, uninterrupted for years, and were reclaiming what was theirs, taking their inherent role in the transformative power of nature.

A wooden house properly built and cared for will last a long time. When we began to dismantle the little house that remained after we stripped away all the additions, stripped the little 40-by-28-foot dwelling back to its bones, we found that much of it was still solid. The wood was dense, the annular rings tight for its slow growth, its colour red-tinted, suggesting that it was Larch, its aroma rich with resin, the smell of it still alive in my nostrils. Nothing wrong with the wood or the construction, only with the care and maintenance, of which it had been deprived.

Our word 'maintain' comes from the Latin *manu* meaning 'with the hand', and *tenere*, 'to hold'. To maintain is to hold with your hand, to be present in the moment, as the French for 'now', *maintenant*, testifies. It suggests 'habitual practice', a continuum of care necessary to keep ourselves and the objects we create from slipping into the wild unkempt. The house that we bought was not touched with love or care, only with patches of panic, vague attempts to insulate it from the effects of its demise, rather than to prevent it. Rats and mice lived under the house, entered at will, brought their kids up in the rotting warmth of its underbelly. They had lived there as long as the humans, made a home of it as much as all the past owners.

For a few months a large bonfire burnt in the tiny paddock where we had built the cabin, the rotten wood of

rafters and beams, window frames and joists all carbonis-
ing in the wind. Slowly, over the months, a new house
rose around the old original one, holding and cradling
it, the grandparent amongst the youthful flourish of new
life. Sentimentality and practicality had prevented us from
totally demolishing everything. The craftsman in me could
not simply sweep away the artistry of others. Once we had
cleaned out all the rot and subsequent additions, the frame
was bare to us, its crannies holding mementos of the time
it was built, a copy of a magazine and a coin. What really
got us to preserve it was partly financial, though in retro-
spect I think that we were probably wrong on that one, but
more particularly how the simple frame had been made.

A hundred years ago, with nails made mainly by hand
and the frame workshop-built, there were no nails holding
the joints together. Rather, every vertical member in the
house, and that would be two hundred or more of them,
had a joint cut into each end, which was housed into a
square-cut hole in the horizontals and held with a wooden
peg. That means that eight hundred or so hand-cut joints
were made, not including those for the cross bracing and
other joinery. Each joint cut and chiselled by hand was
evidence of relationship, of the care needed for the one
shape to fit into the other, of the multiples of unions that
went to create the frame and thus the house.

This was no quick-smacked house of hammer and nail,
rushed to completion for price and deadline. Its quality
was a testament to the pride and skill of the craftsmen
and apprentices whose daily work was this committed
practice. The hammer found its use for the lath and

holding the rafter joints in place, but the building relied on a cut joint, not an added industrial fixing. It relied on the skill of the hand, and the habituation of a lifetime of practice. And none of it was ever to be seen, hidden until we unearthed it under the lime render. Today it is rare to have frames built with proper joinery, and when they are, they are displayed clearly, a statement of vanity, a testimony to skill. Then they were simply the way things were done, not a mark of choice or a definition of the character of the craftworker.

Our new house was built by hand, but with electric saws as well as hand ones, with great skill, yet the need for speed, with certain compromises. Most of the wood came from local plantations, Douglas Fir for the frame and Cedar cladding. All the logs were cut locally as we needed them, and for that I was surprised and grateful. I had a lifetime of practice as a furniture maker, buying wood from hardwood merchants where it was sold for type, not provenance. The Walnut, Ash and Cherry came from the USA, as did much of the Oak. The prized and unusual woods came from further afield. There was little sense of British timber, little supply of it in the woodyards I visited. When we built our house, I began to realise why. Not only were British timber stocks depleted, but the management of hardwood growth in particular was historically very unindustrialised. Few organised plantations existed where the straightness or consistency of the timber could be relied on for the demand that was perceived. It was only in moving down to the depths of the Devon countryside that I could discover a new thinking for myself.

Simon, the carpenter we'd employed to help me build the house, took me down the road to the local mill. I had no idea that we'd be able to get local wood sawn locally from local trees. At that time the mill was run by David, a farmer slightly stooped from a broken spine where a collapsing building had struck him years before. The mill was in a small copse on his land, a great saw able to cut through trees up to 4 feet wide. It was a farmer's business, run with the mentality of make-do, touched with an older thinking. Not much has changed – a few health and safety considerations, a bit of new lifting equipment to make life easier for David and his son Paul, who now runs the business.

All the wood for our house came from here, as has all the timber we have since used to build all the other structures on the land. I was in love with these relationships from the moment I first set foot in the yard. The piles of logs, the great mound of fresh wood dust pouring out of the cutting shed, its resinous and perfumed smell. It was alive with industry, on a scale manageable by a team of three and with sound basic principles underlying it. The offcuts go for local firewood supply, the sawdust for animal bedding, the outer bark-covered planks for fencing. It opened my eyes, reminding me that I had been so blind to the possibilities, and never questioned the way I got my wood.

I soon started buying my own trees: Oak, Ash, Cherry or Chestnut, whatever I was offered from a mile or so around. They were characterful trees, hedgerow or unmanaged woodland grown, marked by the vagaries of wild, lightly touched growth. These would find their way to the mill, and then back to me, planked and dripping wet. Then they were heaved into stacks to dry, awkward and energy-sapping, but at the end of the day I had my own wood supply, drying on the land, and it changed the way I worked. In a word I discovered a relationship with my making through the materials I got, the materials that our location offered. The idea of home, and what it really means was slowly seeping into the bones of everything we did.

———

We moved into our house in 2000, the basics all done, much still to be finished. When we did, I could let all the dust settle and finally appreciate what we were doing. We had arrived on this land a year earlier with a hope and a prayer, torn from where we had been building another house in the hills of Mexico three years before, and dropped into a London hospital. Our daughter, Misha, born with Down's syndrome while we were in Mexico, needed major medical care and we had had to quickly pack our lives into a container and leave the dry heat of the Mexican plains for the chaos of an anxious city life. Nearly two years passed – two years lived on a fine line between life and death. We didn't dare to dream. But at some point after the house was built, very slowly we did dare again, and as we looked out of the windows to the south from our living room, I saw the sea for the first time, and I realised that we had found our way after all. Everything since has been a slow deepening of what it means to be part of a piece of land, to live on it, nurture it and be nurtured by it as best as one can.

Home has become more than a place, more than shelter. Home for me is the manifestation of my material connection to life. It is the point at which wild land and our ingenuity meet, where the soul finds rest and endeavour finds the opportunity for expression. We don't need to build our own house; we just need to interact, to tend a garden or a window box, to cook our own food and make some of the things we require. Once we do these things, we begin to create stories of our connection here, of the meaning of this life, and we root all the more to the earth beneath us.

The land here now supports a hamlet of buildings: the two cabins we rent, our house, an art studio and two workshops, sheds, a bakehouse and a polytunnel. It is home to our working lives as well as to our family. It is a home to other families through nine months of the year, families who come to rest, to take a break out of their busy city lives. It is home to our craft and to that of others who come on our courses.

We have made a life here amongst the trees, shielded and shaded by them, yet at times conflicted over our relationship with them. We have lived here for twenty years, the trees circling us and protecting us, their great crowns reaching over the garden, their shade cloaking the summers, their skeletal beauty marking the winters. They rain their bud husks down on us in the spring, their nuts in the late summer and their leaves in the autumn. At times we must take one down for the sake of safety, and though new ones have been planted, we won't live to see them reaching as far above us as their forebears did.

Chapter Two

Metamorphic

I n her novel *The Winter Vault*, Anne Michaels muses on the Palaeolithic hunter who, in chipping a hand-axe from flint takes care not to damage the fossil of a mollusc imbedded in it. She uses this as a metaphor for the care we need to exercise now, two and a half million years later as the speed of progress accelerates exponentially. I hold this axe in my hand, turning it in the waning light of day, seeking out the mollusc caught on the edge of it. Time has merged it with the chipped flint head, ancient life caught within a tool – the very ancient past within the ancient past, within prehistory and now the present. I am just a passing moment caught within this story, an insignificant blip in aeon time, just a construct of my own mind. The axe head, ancient slice of metamorphic rock, ancient fire blazed into ancient time, reminds me of my tininess.

I cannot help but see mica crystals caught in ancient granite, their tiny reflections barely catching the eye, yet their story much like that of the mollusc and the axe head. Do we care for our environment as we follow the invitation that granite outcrops give us? Mica, feldspar, kaolin, tin, copper and arsenic all owe their presence to granite, as do the buildings, porcelain, pewter, bronze and preservatives that we have made from them.

Devon gives its name to the Devonian period, a span four hundred million years ago during which fish life multiplied exponentially along with plant and tree cover, and the first ammonite molluscs appeared. I take the invitation of the mollusc left within the stone, of nature left within the tool. Ancient granite rock rises in ancient time to the north and west of where I live. It forms the landscape that represents place, home and solidity to me. It rose as hot magma hundreds of millions of years ago, forcing its way through more ancient sedimentary strata. It reveals its eroded presence on Dartmoor to the north and Cornwall's Bodmin Moor to the west, and in pockets further over the Cornish landscape.

The granite tors that rise out of the high ground are sentinels to all the granite that lies beneath the peat earth; the largest granite mass in the UK. They define the landscape here, as the barren moor ascends from the managed farmland beneath it. In warmer times our prehistoric ancestors settled here, the moor was farmed more extensively, and with time stripped of all its ancient growth. The remains of those settlements are the most comprehensive of any throughout the world, and the land has remained barren, heather-strewn moorland ever since.

Standing on the granite outcrops of Haytor or Hound Tor, Laughter Tor or Belever Tor, I can look out over much of South Devon. To my west are scattered various other granite plutons, which pushed through the rock beneath

Metamorphic

Bodmin Moor, around St Austell and Land's End. On a map these plutons, are marked as red blotches amidst the surrounding sedimentary rock. They have been the playground for our family and friends, a wild beat that calls us to them, that enables us to feel some of their wild in us. I have not visited them with an understanding of what resources they have offered humankind or of our extractive relationship with them. I have only seen them and felt them as a part of a landscape that my heart warms to.

Standing on Hound Tor, I turn 360 degrees to take in the landscape around me. My vision falls to the tips of the rocky tors that litter the mottled green, purple-hued, yellow-flecked, rolling hills. From here, they are like rounded building blocks left by a community of giants around the thousands of acres ahead of me. As I lower my field of vision, follow the contours down to the tree-edged crevices etched between hillside slopes, I sense the unseen landscape that lurks below. I remember all my walks along these crevices, down through the contours of this land, where accumulated water tears through the grass and soil on wet Devon days. The moisture clings to the air as the vegetation thickens, held within the leaves of fern and bracken, and the leaf mould gradually accumulating around me.

I am Edmund in the book *The Lion, the Witch and the Wardrobe,* leaving the human world and arriving in a magical universe populated by wild growth and strange creatures. I am at once a pretender in their midst, a human in a prehuman world, a world where time and the natural forces of the planet were the only disturbances. A time where change happened so slowly, through the accumulation of the millions of layers of life and death. The rock is again visible, earth worn away by water and footfall, grey granite exposed and eroding underfoot. The path dips ever steeper to the trickling silence growing around me. Wild nature encroaches on me, and in my imaginal world I feel it inhabiting me, possessing the parts of me never lost to it.

Crossing over the stream and making our way uphill to the back of Haytor, we find the tree cover stops abruptly and bracken-strewn heath replaces it. Up above us the great tor looms, its northern face scuffed and fissured, rubble boulders at its feet, a flattened small plateau running east from it. When we get up to it, we follow it around the side of the tor, carved granite tracks at our feet, the axe head emerging from the rock, the mollusc nowhere to be seen. The transition is sudden. While the wild thrust of ancient magma rises on one side, industry cascades on the other. Great granite chippings littering the ground, the carved granite cart tracks a conduit for the hundreds of thousands of tons of materials that ran downhill on them away from here.

The sounds I hear are of falling hammers, shouting men and the rock straining for release. As the network of canals threaded their way across the country in the 18th century, soon followed by the trundling steam engine, the landscape of locality opened its doors to a world of international trade. This inhospitable place became a quarry, the granite a resource, a material chipped and blasted from ancient rock.

Imagining myself amongst the rolling carts, crashing hammers and yelling workers, I struggle to see the unity of natural forces and our own. Rather, I feel the violation that the landscape was being subjected to as great bastions of time and permanence were split, hewn and extracted. As the writer and archaeologist Jacquetta Hawkes says, humankind has been much more effective than rivers or glaciers at transporting and mixing the surface deposits of the planet. Yet in hindsight, with time and lichens softening the edges, it is almost possible to view the forces of people and nature as part of the same continuum. The granite tracks slowly sink into the grasslands, the quarries fill with water and wildlife, the tips grow over with moss, heather, gorse and time.

The quarried granite rises anew across the country, the quartz and mica in it reflecting light in our cities, buildings reflecting their grandeur and aspirations in it. Each block, faced edge neatly laid in lines, speaks little of the raw granite drilled, chipped and cut, heaved by men from the eroded tor. It speaks little of the trundling carts and workhorses straining through the dusk of a wet Dartmoor day or of the bargemen, sailors, boatbuilders and engineers who enabled its migration and evolution. It says nothing of the millennia before we arrived and evolved to rearrange the order of things. The previous London Bridge, built from Dartmoor granite in the 19th century, now spans the channel at Lake Havasu City in Arizona. Ancient rock, lying dormant for millennia, repurposed, constructed, deconstructed and rising again far from home in the arid landscape of another continent.

———

From my moor-top viewpoint I have come a long way down to the west, following the flow of industry and decomposing granite. Granite as boulder, as hard road edge, building stone, as ancient substrate to much of the West Country. Millions of years ago it rose molten from the earth to set into the geology of this landscape, and all the years later have eroded it back into the textures I see around me. Feldspar, mica, kaolin and quartz jostle within it, localities of provenance altering the mineral compositions. The glint of whites, yellows and greys on worn rock faces speaking of individual components on a chart on the walls of classrooms all round the world.

Over these tors, those of Bodmin and under the busy pattern of our lives, it sits there silently, slowly breaking down to become fluid once again. Time does not stand still, nor does landscape caught within its rhythm, and over the minutiae of an individual life we are blind to the changes, but over thousands of our lifetimes, granite once molten becomes wet clay. From the red-hot to the solid mass of individual substance, and then to the fluid inhibition of kaolin.

This journey I am taking has had many twists and turns to it. Starting on the high solid ground of Dartmoor granite I have ventured downhill with the seeping water. I have found the gullies and channels that funnel that water alongside the washed grit of eroded stone. I have followed that grit to where it leads me, to the stories that lurk in the shadows of those great granite mounds. When the molten lava forced its way up through the Earth's crust millions of years ago to become granite, some of it became changed by the steam permeating it. The feldspar within the granite began to break down and with it the very structure and strength of the rock. The landscape to the west of Dartmoor, over Bodmin Moor, between it and the north-west Cornish coast, was subjected to this transformation on a grand scale.

Two hundred and fifty years ago, much of this area north of St Austell was largely untouched, small-scale farms scattered over the open heathland. Some small pits existed where locals had extracted the sand from the crumbling granite, and the tin-mining industry had exploited the fine white clay that washed out of it to repair cracks in the furnaces. The future promise it held was yet untapped, and the rolling grasslands stood intact. Like a great storm which gently builds, the exploitation of the landscape for the mining of china clay (as it came to be called), started slowly. Yet over a hundred years or more, it built consistently to a high wind that swept away the contours, crevices and fissures of the original landscape, altering them beyond recognition.

I push up a small incline on the edge of a woodland dominated by Rhododendron, hydrangea and some scrub tree growth. I am caught on a gorse bush, my left foot hooked over a branch, and tumble forward slightly as a deafening roar reverberates from just above me. I climb a low gravelly bank, the thick hedge no longer obscuring my view of the ravaged moonscape at my feet. Ahead of me to all sides lies what appears at first as a

37

massive canyon, occupying the full width of my vision and beyond. Textured with the claw marks of diggers and dynamite charges, its white-grey shades echo off into the distance.

Hacked roadways, spiralling down into great depths, yellow trucks alive with industry roaring up the rough tracks with 50 or 60 tons of decomposed granite. Unloaded up at the plant they are soon coming down again through the water cannons and slurry, to the new and ever-changing levels. I make a note of their regularity, and it works out that there are twenty trucks passing me every hour. I feel compelled to view, yet my horizon is choked with the uncomfortable reality in front of me. As I spin my viewpoint a little, raising my head to the left and right above the pit, there was no landscape to see. All that stood before me for miles were conical mounds of waste, old tips grown over with sparse vegetation, great hangers and steel slurry tanks. The industrial landscape of the last 250 years, engineered by humans, is fissured deeper amidst the conical slag tips, mica dams and yawning chasms. This pit is by no means unique; it's one of hundreds of old and operational pits that take up a massive area of 30 square miles of the moorland of central Cornwall.

Between the pits and mountains of debris run rows of arterial villages, community halls and churches. These grew as the mining industry here developed from the late 18th century after the discovery by William Cookworthy of china clay. After some experimentation he discovered a recipe by which locally occurring granite clay could be mixed with local china stone to make a type of porcelain similar in quality to that used on expensive Chinese imports. Unable to get good-enough results himself, Cookworthy's enterprise was taken over by Josiah Wedgwood and other Staffordshire manufacturers. The industry grew massively over the next hundred years, and Cornish china clay became the third-largest mineral product of the UK, after North Sea gas and oil.

As I left the viewing platform, I headed back downhill along a small stream, its gentle flow running clearly over a bed of grey-blue granite sand. Trees in young whippy growth covered the ground over which I walked. This side of the pit I was on was protected land, and the trees' growth disguised the land's disfigurement. This stream would have flowed with clay slurry, a pumping station where I had been standing drawing it up from the pit floor deep below. As it flowed down away from me, a line of steel bars riding on rollers would have clacked their way backward and forward alongside me, powered from a bigger stream on the valley floor below.

As I followed them down, a great waterwheel rose ahead, turning laboriously as if at any moment it would lose momentum and stop. Yet as the weighted arm moved slowly to the bottom of its swing, the whole wheel gained momentum once more and spun freely. With each stroke it would have drawn 20 or so cubic feet of slurry from the pit floor and let it loose to run along the little stream I had recently been on. Away from the wheel and down to my left, a row of granite buildings ran a great breadth in front of me, and moving between them and the waterwheel,

I came to a series of shallow tanks. Some of these were divided into channels, others just open. A small hut, like others I had passed, stood at their side, a small waterwheel within it.

I paused to take stock. I had walked downhill some half mile from a working pit in the middle of a great wasteland, through a small woods and now into a great old industrial enclave. I had walked from the 21st century and into the early 19th century. I had walked from single men sitting in the heated cabs of great machines, and down into a memory of many individuals with pickaxes, mattocks, crucible-like scoops and shovels, all labouring to do in a year the work that man in the truck could do in just a matter of days. Till the early 20th century the work relied on hard labour and gravity. Water running downhill powered water and slurry to be pumped uphill, and gravity took them to the tanks. There the grit was separated from the finer particles of clay, and the damp kaolin was brought into the great sheds for drying and finally packaging. From here it went by horse and cart to the docks at Charlestown for shipping abroad. The remaining undecomposed stone was loaded into carts and drawn up, first by water power and later by steam, to be dumped at the top of great conical tips.

The small river from which the great 35-foot lower wheel at the mine gets its power is known as the White River, and it runs along the valley floor from Roche twenty miles or so to St Austell. In the 19th century all the water washing through the cleaning tanks for the clay drained down to that river and from there down to the sea. Fine clay and sand held in suspension stained the water white and killed all the life within it. The sand and grit lined the riverbed and was washed down to the beaches at the end of its journey to the sea. Within a stone's throw of this mine there were two others, one of them now a reservoir from which to pump some of the 60 million gallons of water used every day by the clay industry. The road I had travelled here on would once have been lined with the chimneys of the clay dries, and the river would have been thick with the industry of them.

This mine at Wheal Martyn is one preserved of the hundreds that there were, and in it is evidence of the community of makers supported within the industry. The great buildings and tanks I had seen on the horizon earlier were automated processing plants to which the unprocessed clay was taken. Everything I saw laid out comprehensibly around me as I looked at the 19th-century workings happened invisibly and probably quite incomprehensibly within those modern warehouses. There would have been carpenters here to maintain the carts, mill races and all the other woodwork. There would have been coopers to make the thousands of Baltic Pine barrels into which the clay would be packed. There would have been blacksmiths and local foundries to build and maintain the waterwheels and other ironwork. The trades would have been alive here; the crafts had meaning as part of essential mechanisms of production and trade.

I admire this, and sense the community and identity within it, and how difficult change must have been. Yet the image of that lunar landscape of torn moor, white wounds

gaping to the sky, reminds me poignantly that these iden-
tities and skill sets can en masse and unchecked do great
harm. As I wander the remainder of the works, I feel in
the core of me the pathos of this, that in our nature we
tear at the very nature of ourselves. Yet I am in wonder of
what they achieved with water and ingenuity. The stream
that drove the wheel had brought water off higher ground,
and as that wheel pumped the water up to where it was
needed, it could drain back with gravity and the help of
other wheels to drive the cleaning and processing of the
clay. Before the advent of high-pressure hoses, the water
pumped up had been allowed to drain back into the pit,
over the tools of the digging men. It washed the finer
particles down to the pump and started the whole process
of transformation.

From water it came and to the water it would return.
The clay would be loaded 3 tons at a time onto carts pulled
by three horses, two carts travelling at a time the five miles
or so to Charlestown Harbour, two extra horses in tow.
When the incline got too steep, the two extra horses
would be used on one cart, while the other driver waited.
Then the first driver would bring them back down the hill
so they could be hitched to the second cart. Upon arriving
at the harbour the clay, if already processed, would be
taken straight to the linhay for storage. Or, if it had come
processed but not completely dry, it would get taken to the
clay dry to complete its journey of transformation. From
here, after a few days of drying, it would be pushed in carts
down tunnels to the linhay for storage as it awaited the
arrival of the ships.

Metamorphic

———

I find myself in a café overlooking the beach near St Agnes, the industry of the day and days past spread before me. The car park is slowly filling, the lifeguards are setting up their flags, and there are just a couple of surfers testing themselves in the white foam. Dogs criss-cross the sand, opening the width of the beach to my scrutiny. Here in the 21st century, this is one of hundreds of surfing beaches on the Cornish north coast, fine grey-beige sand spreading far out into the low tides. Broken rock lies in piles under fallen cliff faces, and lone stacks stand forlornly amongst them. To my left the rock piles are notably different, many square-edged and paler in colour. Cut granite lying tossed amongst more natural forms of darker stones, it speaks of something more than crashing waves and plastic surfboards cutting through them. It speaks of a time gone, which laid the foundations for this view.

Geology in Cornwall leads the way, as usual. Lodes bearing tin and copper cut through the rock strata, a memory of ancient geology, a molten bubbling of granite releasing mineral ores up into the crust. The cut blocks are all that remain from the last of five harbours built here at the beach in front of me between the 17th and 19th centuries. Ships used to transport the tin and copper ore hidden in the lode-bearing rock all around this area. The cliffs above are pockmarked with rectangular slots, a memory of the gantries that would have hauled the tin onto the boats, and the lime, coal and timber off them.

After a chat with the young man working in the café, I find myself staring uncomprehendingly at a 3D computer-generated video. I am transported through an arterial maze, a sequence of vertical lines punctuating the black landscape, horizontal and oblique veins moving off them at intervals, the whole morphing continually with some hidden rhythm. Each vertical line gives birth to a whole series of branches, which seed more minor branches, the whole forming an inverted tree. The branches of these trees overlap those of the next vertical line, creating a mind-boggling and visually arresting pattern.

The fact that I'm virtually travelling through a 3D representation of the underground tunnels of tin and copper mines makes the whole experience surreal. The vertical lines are mineshafts, each one so close to the next: another prospector of the 18th or 19th century, another hope, another fortune made or lost. The horizontal and oblique lines follow the lodes to search for the valuable deposits. Above the whole, up on the surface sits a repetitive pattern of old engine houses, some preserved only as a name or memory, but they once followed one another in close proximity up and down this coastline.

A few weeks later I get to see a section of this 3D image showing some of the mining area of north-west Cornwall. This time it is a great model housed in a large room, Perspex and coloured thread replacing the digital markings on the computer screen. I am at Geevor Tin Mine, the last working tin mine in Cornwall, and this model was the way that the engineers could understand exactly how the industry of the last two hundred years had shaped and perforated the ground beneath them. They built it to plan new exploration, to further fissure the earth. As I looked at it in its physical form, I was able to see just how extraordinary the results of their industry were.

Walking up onto the cliffside from the café a little later, I look back onto the valley running down to the harbour. I can immediately see eight preserved chimneys within a very small area of no more than a square kilometre. The footpath winds its way uphill, its edges lined with gorse and heather, the land uphill from it climbing steeply into piles of barren loose rubble, slag once dug from deep underground and now piled forlornly here. As I reach a high point, I see over to the next valley and three more chimneys deep within it, and one old engine room, its chimney no longer standing. I imagine the industry here, the coal smoke firing those great steam pumps, the labourers above and below ground, the scarred hillsides and polluted watercourses, their beds and the beaches they flowed into stained grey with the rejected rock. The industry I stood amidst earlier was one of tourism, the cafés busy on the beach below, the surfers feeling the thrill of the wild, couples relaxed on benches overlooking the spectacular scenery. The industrial coastline of yesteryear was rebranded as a cultural heritage, and wilful nature had done its best to repossess it.

Geevor mine closed in 1985 after the price of tin, finally unfettered of strict control mechanisms, plummeted. A general demise that had started in the late 19th century ended with the death of the industry. What had for a few hundred years been such a dominant part of the Cornish landscape and heritage went silent. The chimney stacks became idle, the beaches slowly changed colour, and men went elsewhere looking for jobs. The collapsed harbour was left, the brick kilns, warehouses and associated industries fell into disrepair, to be repurposed one day as holiday homes. As I stand looking over two valleys, I imagine the lodes that formed under heat and pressure three hundred million years before as molten granite pushing through the shale surface released mineral deposits previously inaccessible. Lead was pushed to the surface first, copper below it and tin below that. As the rivers eroded they mined themselves through these lodes and left the minerals exposed and readily accessible. The softer metals were easily worked, but it wasn't till four thousand years ago that enough heat could be generated to fuse tin with copper and create bronze, and thus the Bronze Age in this part of the world was born.

I walk down into the smaller valley with images of figures panning and pounding in the water flowing along it. Coming to the valley floor, I walk up along the stream past the ruined engine house and through piles of discarded stones left by bal maidens hundreds of years before. I stand in a small yard of tin-roofed wooden sheds, the sound of a waterwheel coming from behind a fence. This is the last working tin-production enterprise not only in Cornwall, but in fact the whole of the UK. Mark Wills, whose father, Colin, bought the land back in the late 1980s, lives on site and has repurposed an old industrial mine site into a craft initiative. Glass boxes on the front counter at the entrance display the cast fruits of his labour, bright tin jewellery and candlesticks made from the tin that they process here.

Metamorphic

Old photos show the site in full industrial action in the 1900s, waterwheel turning, engine house pumping and big sheds alive with the industry of men. The tin produced would have gone off unprocessed on ships alongside that of all the other hundreds of mines. Yet now they don't mine the black tin ore here, but rather have a licence to collect fallen tin-bearing rock from the beaches. Mark reads the landscape and character of stone as he collects what he needs for smelting the year's supply. He is of four generations of miners, and uses techniques that predate all the sophisticated engineering of the deep mines. He is simply collecting the rock that eroded off the exposed lodes as it has done since time immemorial, and he is not in competition with anyone over it.

The parking area is home to Mark's Land Rover, and nothing else. The pickup back is textured with stone gravel and dust. The area beyond the fence is still, only the rhythmic beat of the waterwheel a memory to what had once been. Also, a testament to the new industry, tourism, that drove Mark and his father to reconstruct it when it became clear that they could not make a living out of collecting the ore alone.

The piles of stone have sat there for years, and there is little to see of the ore Mark collects from the beach every winter. The rhythms of the season dictate the rhythm of his work, as winter storms reveal new tin-bearing rock, and his pile of rocks swells over the cold months out of sight behind a hedge. The rhythm of the wheel takes on a new sound as the rock is loaded into the stamps, which the water's power raises and lets fall. Slowly rock becomes gravel, and gravel a fine powder. The rhythm of the seasons and that of the pounding sounds are driven by the power of water, revealing, loosening and crushing the cliff face into a commodity.

Yet this is more than a tourist venture, more than some shadow left from a distant past. As I talk to Mark, I sense his attachment to the rock at his feet, to the heritage he works within. His desire to learn how to process the tin ore into pure tin, to restore the stamps, and to create a market for his products stems directly from his connection to his heritage and locality. While other mines like Geevor stand derelict because tin prices are too low to sustain them, Blue Hills Tin soldiers on, because it has kept tin within it, made it a material of its own sustainability. Geevor and the other mines produced tin on a grand scale as an international commodity, its price relative to worldwide production. Blue Hills produces tin only as a material for what it makes. Its scale is tiny, and as such is a model of miniature sustainable practice. It is a small remnant of a massive industry, but its scale is not important. What is important is the story it continues to tell.

As Mark collects, transports and transforms rock into tin, he processes the very essence of our life here. He uses heat and power to convert one material into another, and the end product he creates is largely irrelevant. In many ways he has many end products. He has the stone he collects, the tin he extracts from it, the objects he makes using the tin and the satisfaction of the tourists who learn about the process when buying his objects.

I agree with Jacquetta Hawkes and her comments on the 'industriousness' of humanity. On one of my expeditions to Cornwall, I ventured down the shafts of the Poldark Mine (formerly known as Wheal Roots). I was able to imagine how it was in the 18th century, before steam power, before the might of the industrial machine was whirring at fossil-fuel speed. Men pulling on wooden wheels, winding chain up wooden pipes, jammed rags pulling up water, emptying it from the pit. Four-foot-high tunnels devoid of light, candles sitting precariously on resin-soaked felt hats, little air, water pouring in at speed. Thirty years was the average life expectancy of the men, twenty-two years of servitude down the shafts, twenty-two years of shattering strain. The women in the community often had three husbands, had been widowed as many times, had suffered repetitive loss. They toiled at the surface of the mine, the bal, watching their own kids go down at the age of eight. The industry of the human race comes at a price: the toil of a person's body to extract profit from the natural world has worn itself to the bone. The hollowed-out lodes deep under the Cornish granite, stripped of their ancient mineral history, are to most eyes as silent at the dead youth who struck and thrust their way down into them.

The tunnels and shaft at Poldark Mine are pre-steam, of a time when picks and shovels and eight-year-old boys struck and dug the granite out from under the earth. They predate explosive charges, the Cornish developments of fuses and pumps, of technologies that might have assisted them. They don't go very deep, but 100 feet or so seems deep enough right through the hard granite core. The granite was discarded; that it had no value in itself reflects the focus on the tin, on the coin, the value hidden in the rock. As I bent my way through the tunnels, the guide remarked on the line cut through the rock 2 feet lower, that this was the original level, that the miners, though shorter than me, had to stoop and crawl through these tunnels where they lived six days a week. No wonder they believed in the knockers, the mine spirits whose knock they heard alongside their own.

When life was so hard, when toxic arsenic or silicon cut it short beyond the labour, belief was something to treasure. Arsenic occurred alongside the tin, lone chimneys near some pits being old arsenic refineries and reminders of their coexistence. The silica in the granite, when drilled or pulverised to powder caused silicosis in the lungs of the miners. They are dead, and cannot speak, but the contaminated soil does, unable to support plant life, especially on copper slag, and reduced to a few mosses and liverworts that can survive on it.

The water ran noisily under our feet, a pump sounding occasionally, sucking out tens of thousands of gallons of water every day, leaving us relatively dry where without it we would have been drowning. Without that water, there would have been no mineral-bearing lodes, and yet the water constantly seeping into the mine threatened to stop the miners from penetrating to any depth. The Cornish beam engine pumping houses that define the landscape in Cornwall are a reminder of the presence and power of that water. There were once more than three hundred mines in Cornwall, many with more than one pump house. The coal shipped back from Wales after the copper ore was taken to the smelters would have driven these great pumps and replaced the water- and man-powered pumps of earlier times. There is one of these great pumps at Poldark Mine, 50 feet or so high, cast steel with bronze bearings, great pistons and beams travelling down the shaft hundreds of feet under the earth.

Outside the mine was a low and innocuous outcrop of rock, or mortar stone as it is known. It predates those great 19th-century pumps by nearly two thousand years, from the time when panners were searching for surface tin.

I feel the disturbance in the earth. I feel the disturbance under my feet. I feel the disturbance in me. I have an image of the emptied lodes, underground scars of our search for one mineral. I feel into the hollow within the granite, and imagine my veins emptied of blood, my nerves stripped of their communication networks, my muscles stripped of their sinewed articulation. I would be only a shell, a skeleton stripped of its structure, of the soft organic that keeps it in life.

The granite core, igneous outburst, is left hollowed, too, hard rock tunnelled out, letting humans in. Shafts penetrating vertically, connecting to routes of exploration, hundreds of miles of tunnels underground, under the sea, underwater. As if I, too, were spread out on an operating table, with industry and technology boring into me. Needles and cameras, scalpels cutting and slicing. This I can feel. The waters in me pumped out to facilitate the exploration, the quiet whirring of machines to keep the forces of nature temporarily at bay and allow humanity its moment of glory.

Material

It strikes me that we search for impermeability, for immortality. Diamonds and gold, silver, tin and copper, hidden minerals that become visible objects of status and longevity. They are associated with the length of our marriages, with our value reflected through theirs, with distinction and privilege. Wood and bone and materials on the Earth's face share little of that association, and they rot quickly back to their origins. They are the compost of our future, and will disappear quickly from their human-wrought form. Though they command no real value in our civilised society, their inherent value is immeasurable. They are reminders of our own mortality, and their passing leaves no scars, for where the trees grew, others will grow anew, the ground enriched.

Chapter Three

Selvedge

U sed by kids to mimic gun or sword, used to hit balls to test prowess or raised whimsically as a spear, in play the stick is a symbol of power separated by a gossamer-thin veil from the darker side of adult behaviour. It is often associated with masculinity: the 'boyness' of boys, and how they 'naturally' gravitate to certain forms of play. Stick, bat, weapon and stone, ball, missile are two sequences that seem difficult to escape. Inanimate objects held in human hands become either toy or weapon. This is often used as a shorthand for human development, the idea that playground behaviour is being practised in preparation for the world outside the playground.

The premise is that the stick and stone were the first interface of our ancestors, the way that earliest Homo sapiens differentiated themselves from the rest of the animal world. So the story goes that inanimate objects became weapons with which to exert power over others. Humanity's mastery of them has become increasingly refined so that these objects no longer look like the sticks and stones that were originally picked and lightly fashioned from the primeval forest floor. In only a few hundred years, they have become extraordinarily sophisticated weapons of power with great destructive potential.

Tools have a role within this overarching narrative, too, in the articulation of stick and stone, of spear and missile for hunting, driven by the hard edge of resilient materials. We have all played rock paper scissors, I imagine. I was always so surprised that paper had the power over the rock that it did. There was something poetic about this, something that transcended so much of what was otherwise fundamental thinking. That a material so thin and delicate could overwhelm something so resilient, heavy and powerful as a rock never failed to bewilder and entrance me. The three protagonists in the drama are the hard and heavy, the soft and expansive, and the sharp. So scissors cut paper, which when uncut smothers rock and takes away its power.

It seems that they are in perfect order for the timescale of human evolution. The stone occurring in its raw form, lying cracked, worn and separated on the ground was picked and roughly fashioned with other rocks. It could be used to crudely pound or shear fibre from wood, bark, grasses or skins. These fibres – woven, pounded, saturated or heated – were used to fashion clothing or baskets or parchment-like sheets. The hard helped create the soft, and the soft helped contain a life, to protect and keep warm or to collect food and radically multiply the holding potential of the hands.

The scissors of the game are clearly a very late development of the tool, a mechanical adaptation of the steel blade that has been around for only a few thousand years.

Yet its mechanistic imperative, or at least its antecedent – the knife edge, arrow head, sharp point hafted to a stick – has driven much of our thinking.

There is an essay by Ursula K. Le Guin called 'The Carrier Bag Theory of Fiction'. Her theories tie in with my own musings, and help create a more rounded vision of tool development and use, and of the imperatives that may have driven our relationships with making. In it she imagines the seductiveness of the stick and stone narrative, of the hero prototypes based around meat eating and hunting practices. She seeks to challenge how prevailing narratives end up dictating regular patterns of consciousness and of fictional and factual storytelling. Working on established research that between 65 and 80 percent of our ancestors' diet was vegetable-based, she imagines the bag as great an imperative as the spear, or greater. That the hunt for vegetable sustenance and the development of humanity were contained within the container that contained the fruits of the day's labour.

Her main point, though, is how the acceptance of a particular narrative can then affect all the stories we tell. By telling stories based on the assumption that it was the spear and not the carrier bag by which human development was articulated, we set in motion later stories coming from an aggression-centric, male-centric, meat-centric perspective. How different would our narratives be had they been based on the idea of the bag or basket, of holding, of containment and of the different relationships required with nature that arose. As far as I know there are no cave paintings elevating the basket or woven cloth, but there is this idea that woven plant fibre has a strength, resilience and contribution far greater than the disappearance of it from the historical archive would allow us to believe. There is relatively little archaeological evidence of woven plant fibre in comparison with the extensive catalogue of stone and metal weapons and tools.

———

I am out in the car again to explore these ideas, finding makers who work with pliable fibre, who follow the lineage of the bag, who work within the relationship of grown and woven fibre to make baskets.

I've always loved Ordnance Survey maps; never really thought much about the name, only that it hinted at their dependability and reminded me of the promise they held over my exploration of the wild terrain outside of my safe domain. The first name for the Ministry of Defence was the Board of Ordnance, and with the Scottish rebellions of the 18th century on the back of the French Revolution, there was a perceived need to build defences from invasion, and to do so to construct an accurate mapping of the country. This started in Scotland and took over a hundred years to complete.

Though first meant for defence, the surveys quickly assumed many other usages, including industrial development and the taxation of land. The first maps sold in the early 19th century cost the equivalent of one to three weeks' average wages and would have been available only to wealthy landowners or organisations keen to get a bird's-eye view of their possessions. The map I had spread before me on a March morning as I planned out my journey for the day had cost me £8.99 and was principally designed for leisure activities and sold to walkers or tourists keen to explore the environment. Perhaps this is a form of modern warfare, the invasion by strangers to a certain landscape.

I am thankful for the map's existence and the ability it gives me to imagine the landscape that I'm part of. There is a beauty to standing in a place, seeing the horizon, or the hills or vegetation that blocks our view of it, the texture of the ground at our feet, or the constructions that are visible to us. Yet it is a narrow prospect; it tells us little of what is on the other side of a hedge, or of the dense copse ahead of us. In Devon the high hedgerows, though beautiful, separate us from the landscape around.

I was looking to see where I was going that afternoon for an appointment with a basketmaker, Lin Lovekin. I wanted to have the tactility of a map. I found where she was, no name for her house, but the name of the hamlet she had given me clear on the map. As I identified which lane I would take, I noticed the scars and marks on the map, the words AIR VENT, MINE or METAL appearing repeatedly. A quick search revealed that her house stood right over one of the most important 19th-century tin mines, Wheal Vor. So here I was, on an innocent search to find where a maker lives, wanting to interview her to discover something of her relationship with natural and perishable materials, and, lo and behold, I had become connected to the idea of materials once again. Under the ground beneath her postcode, twelve hundred men had toiled to extract copper and tin from the earth. Behind them, after a couple of hundred years, they had left slag heaps and fallow ground, on which little could now grow but scrub, and etched into the printed pages of maps those scars were furrowed for posterity. While above the ground, materials were grown and woven into baskets, below, the land was ravaged by stick and stone, tools and machines – progress.

When I did find my way there after lunch in a nearby pub – my waitress ignorant to the tin-mining history of the area, the road peppered with the chimneys of now disused mines – I found so much life amidst the detritus of past industry. The path snaked off the road, potholed and twisting amongst the scrub on either side. I couldn't see any fields for the branches of gnarly shrub-like trees, and it felt as if I were in a time warp, entering another temporal plane. The land beneath the growth was flung about, unnaturally moundy and uneven.

Keeping literal to Lin's directions I kept ever to the left, circuiting almost back on myself until I drove as instructed into a small chaotic yard with an open barn and timber piled by splitting blocks. I like it when I am disarmed, when a foil is raised to my face and I am challenged to step away. Nothing in this scene spoke of exactly what lay behind the tangle of logs and old sheds, nothing at all. Yet I knew there was magic here, in the alchemy of materials beneath and above the ground, in the practice of making and the idea of home.

Stepping out a little hesitantly I looked around to see a hand raised at some distance from a shed beyond the short field of my vision. A voice called over to come that way, a path that looked to me as if it led over the chickens' electric enclosure. Yet trusting the arm, voice and now face of Lin, I found my way round the fence and into a bright hut that turned out to be her studio. A stranger arrives in an intimate space to find a maker just recently present only to their own process. There is a level of intimacy here that needs utter respect, a responsibility to tread lightly enough to allow a shift to a shared space. The studio was newly built by James, Lin's woodworking partner, its walls clad in speckled board, its frame honest and appropriate. Light came in through translucent sheets, and the space was alive with the clippings of Willow and the great open baskets on which Lin was working, which turned out to be giant lampshades.

Lin worked as we talked, and I was immediately struck by our ability to do so without any noise to interrupt us. Her movements were purposeful as she clipped away the excess stems from the finished forms, and her focus was with both them and our interaction. We had met as makers with a shared language. We talked of process, of making a living, of lifestyle, of clients, of joy and of toil. We talked our way through five clipped baskets, through logs burning down on the fire and right through any timidity that might have stood between us.

I was grateful to be met with such trust and to find such communality. I was grateful to be privy to the intimacy of another maker's practice, to their sacred space, that liminal place between the materiality of the physical and the ethe-reality of the spiritual. We talked of this also, of the senses and what lies beyond form, of the feel of a thing or of an emotion, of smell and sound and other factors that enter the subliminal world of a maker's process.

The smell, something of the eucalyptus about it, sweet yet acrid, lingers in my memory, much as a vivid image does, or the memory of a conversation. It strikes me that texture for a maker is all of this: the smell, sound and touch of a thing. The smell soothed me, and reminded me of all the people who have entered my workshop and commented on how soothed they are by the smell they encounter. As I watched Lin's hands work with dexterity and purpose, I realised that a maker's art comes largely through feel, that in the moment-by-moment interaction with a material, its resistance, or surrender, guides our hands as they dance around the natural properties that we are in dialogue with.

It was a wonder to see the 4-foot-high forms that she was working on, and imagine how they had risen from the lashed bundles of sticks leaning against the wall behind her. Colours – reds, oranges and greens – varying shades of natural bark, and then the bleached white Willow stripped of its bark. The colours bunched at her side had come from the Willow beds in her garden, each colour a variety, cut while dormant, the sap still down, buds formed but no leaves.

The plants look a little sad now, pollarded back to a few inches off last year's growth, all standing upright in regimented rows, waiting for the spring to reinitiate their growth. Lin and Jamie's garden is a 3-acre plot populated with their house and various outbuildings, their workshops, Willow storage and the goats' home. Goats and chickens, enclosed by electric fences, take up more space, as do the vegetable beds, Willow and coppiced Ash for firewood and green woodwork. Everything is purposeful, and much the same attitude that imbues their perspective on making. There seems to be little ego at play here, but just a desire to engage physically and soulfully with the material that so abundantly grows around them, alive with their husbandry.

When Jamie arrived after Lin and I had talked for some time, the conversation became a little headier, the two of us finding that we had enjoyed similar books, and pursued similar trains of thought. At one point Lin showed her exasperation, declared that it was doing rather than discussing ideas that appealed to her more. We came back full circle to meditating on what lies beyond the physical engagement, what it is that defines a maker, pulls them to the action of making rather than the result.

Their home, its history and the growth on the land confirms the impermanence of everything, challenges the idea that an object is something fixed, rather than just another stage on the road to decomposition. Lin spoke of her baskets returning to the earth after their purpose was complete, and there is something very beautiful in that. It reminds me of Art Carpenter's statement about the buildings on his land gradually being repurposed by nature, about the idea that nothing is permanent. That the impermanence of what we create is merely a mirror of ours here on Earth, that one day we will all be repurposed in service to the whole. It is a conceit to think of ourselves or anything we create as permanent, yet we are obsessed by the idea that we will live on, and that the objects we create will be indestructible.

It seems that we can't accept the idea of degrading, or of death. On buying timber we are assured that it will last many years, that there is a new product that will last even longer, yet we have never really known why! That there was arsenic or copper nitrate in these products was something of little concern or never known. Where had it come from, what harm had it done and what harm might it do when finally rotting into the earth?

Under the house where Lin grows her baskets, copper was once mined alongside the tin, and there was probably arsenic present in the rock, and if not, there were certainly arsenic mines nearby. It is ironic that she is making with a view to the honourable function and dissolution of her baskets in the very place from which minerals were extracted on whom a narrative of immortality was placed. What we extract from below seems to convey permanence, after sheltering there for the millennia before people extracted and repurposed it. While the materials above the ground, growing and regrowing on yearly cycles in a constant rhythm of life and death will, on their final departure, return to the ground to give life. So we in our conceit use industrialised thinking to prevent their decay, and ensure that when they finally do succumb, that they will leave their underground toxicity on the soil so diligently gifted by the decaying leafscape.

My first trip to see Lin was in March, the cool of winter still choking growth, the pruned Willow fallow and to my eyes empty of promise. After a six-month gap I was back again in September, the track even bumpier, the brambles wildly grown over the slag-strewn land on the edges.

Walking with Lin in the garden after a meal supplied from her beds, I spotted the Willow, head height and abundant in the gloaming. She smiled with pride when I commented on it and on the beauty of the land, the abundance she and Jamie have created here amongst the fissured texture of past industry. She had opened the goats' enclosure, and the dog, excited in its perceived role as shepherdess (lots of excited barking and tail wagging while it ran round in circles) inadvertently chased them to their pen. After they were safely locked in and fed, I had learnt a little more about life here, about what home meant to them, about belonging to a community of land, animals and family. I was touched again as I had been six months previously by this synergy of a working, productive and meaningful life.

When I left a little later, the proud bearer of a beautiful driftwood-handled basket of Lin's, it held some of this life within it: Willow from the land, honey from Jamie's bees, soap made using the goats' milk, and all the purpose and commitment that living a materially meaningful life takes.

It is not such a great leap from Lin and Jamie's life to the stories I have read of the indigenous peoples of the American Northwest coast. The communality of their lives was held within a deep connection to their natural environment, yet also quite specifically through their relationship to the Cedar tree. Through it and the materials they extracted from it, these tribes clothed themselves, made tools for fishing and hunting, boxes for food storage, the houses they lived and gathered in, and much else. It provided directly or indirectly for their basic needs of shelter, food and community. The Cedar tree was in a sense their mother, *mater,* the very source of their material existence. They had very limited choices within the forest environment of their home. Yet they also possessed everything they needed by developing skills, tools and a deep relationship with place.

With only primitive tools available to them, napped stone axe heads bound with Cedar wood fibre to Cedar wood handles, they fashioned beautiful and functional objects from every part of the tree. The bark was stripped and woven into baskets, rope or fishing nets. Fallen trees were repurposed into canoes, which were hollowed out with the aid of controlled fire and stone axes and adzes, or into communal post-and-beam houses. Living trees were allowed to continue growth while planks were carefully split out from one side of their trunk. These were used as cladding for houses and covering for roofs, or they were cleverly carved and steamed so that they could be folded along their length, and then stitched together on the meeting end grain to form boxes for storage.

All the objects were decorated with stitched, carved or burnt symbols that spoke of the tribes' profound connection to nature, place and to the sacred Cedar. The spirit of nature and the gods whose narratives contained it were summoned into the very fibre of their made world. To them the Cedar tree was known as 'long life maker' or 'rich woman maker' because it provided everything needed for a long and full life. They had no doubt as we do today of their connection to the earth under their feet and all that grew from it, and with it they wove the selvedge of their lives and community, the tightly woven form of interconnection and cooperation through which they lived a life in harmony with the natural world.

That energy, once so abundant within community relationship, is nowadays transliterated into a narrower communality within and around makers' lives, and the relationships they have within the community of other makers. The makers that I've visited in the writing of this book pursue many different crafts, work in different parts of the country and have come to my attention for many varied reasons. I have not sought well-known craftspeople, nor have I trawled through directories or made exhaustive lists. I have simply followed the threads as they have presented themselves to me.

A chance conversation with a stranger, an enquiry into an object that has taken my fancy or remembering someone whom I have been familiar with for some time – it was like this with Hilary Burns, a well-known local basketmaker who has done much to give attention to the once ubiquitous and necessary craft. She has been on the periphery of my vision for the last twenty years, images of her baskets on the white brick walls of a local café, references to her, my son exhibiting with her, all contributing to a gentle and gradual process of getting to know her.

Hilary is responsible for one of the best definitions of 'material' that I have yet heard. When on going to visit her recently she answered my question of what the word meant to her by saying: 'What you can lay your hands on.' I had found my way across country to her beautiful thatched home a little earlier that day. The sun was bright, sky blue and the house appeared as if a picture postcard to a foreign tourist, perched on the road edge, a sentinel of bygone times.

I was met by the sensory and tactile experience of walking on pebbles, strewn on the floor as if I were on a beach, imbedded directly into the red earth, worn down and polished by constant footfall. I assumed they had been newly laid, but learnt from Hilary that they were original to the cottage. Thatched roof, cob walls and pebble floor all gathered from the landscape around. Reeds and clay and beach pebbles, things that the original owners could easily 'lay their hands on', materials which were present in the landscape and to which need gravitated, and with it, skills developed. Hard stones, soft clay and the pliability and strength of vegetative fibre. Small windows, cool interior, small panes of glass all pointing to the more limited choice available to the local builders, the low ceilings and doorways forcing me to stoop.

As Hilary set about making tea, her friend Geraldine, also a basketmaker, was out the back in what had been a scullery going through bundles of different-coloured Willow, setting the weaker pieces aside and bunching the remaining ones neatly together. We chatted briefly about basketry, and Hilary brought out some baskets that Geraldine had been weaving from stainless steel braided wire. The tension inherent in the material allowed a pattern of loops to form repetitively through it, texturing the surface and giving a unique quality to the uniformity of the wire. I couldn't help but play with them myself, obliging as they were to shift shape with the slightest pressure, reminding me of the wire craft that was common in the town where we had once lived in Mexico.

Selvedge

The bolts of Willow standing in pots, colour-coded by nature, were rooted to the traditions of weaving that had grown out of them, out of the Willow beds. This manufactured stainless steel wire, imported from somewhere or another, full of its own promise and narrative, stood out sharply in their shadow. Beautiful as the craft and forms were that had made them, their presence signified a rupturing of time, a sharp dissonance to the raw materials which were so abundant around me in the structure of the cottage.

Geraldine left, leaving Hilary and me chatting over teacups at the kitchen table. I learnt of her roots in Zambia, of art school and working at a knitting mill in South Africa, of unpicking patterns to understand how to reassemble them, and how this imbedded in her an interest in how fabrics were constructed. Aged twenty-one, she came to the UK and was soon helping students to unravel their problems and work out how to construct their ideas. Slowly she was pulled towards basketmaking, as an extension to weaving and because it was not something that a machine could replicate effectively. The nature of the raw material, and the three-dimensional shape of the basket itself made machine fabrication difficult. This three-dimensionality appealed to her, her interest in how something went together exercised through the challenges of working, bending and weaving Willow into containers.

The same ideas of warp and weft that hold a piece of fabric together create in a basket a strength and durability that can take great loading and impact. She talks of the selvedge, the 'self edge', the way that a weaver finishes the edges of fabric, the skill needed to do this neatly and the strength that it imparts to the fabric. My own little experience of weaving had already acquainted me a little with this, and how difficult it is. A weekend course I took within the writing of this book with a local weaver once again brought me face-to-face with the intrinsic complexities of process that can so easily be taken for granted in the mere observation of a made piece. The piece of fabric I finished with, striated by colours that I had chosen from a 1970s adolescence, was contained by a selvedge twisting and contorting along its length. It would clearly take more than a weekend for me to define the edges of my own potential.

As I listened to Hilary from a position of inexperience and naivety of the craft, I saw that this was vital to her, this selvedge, the finishing around the woven material, and how in baskets it becomes so important. It is where much of the strength lies, and where the strength of the maker's hands gets tested. The conversation later touched on novices learning the craft, and how this process of twisting and tightening the Willow around the edges of the basket was really challenging to the undeveloped muscles of their fingers and hands. The strength of the basket is reliant on the strength of the hands, the atunement of the maker to the utility of the basket.

Handmade baskets are things of great beauty, and have now become lifestyle accessories or pieces of art. Replaced by the plastic bag or the cardboard box as containers for shopping with, they no longer have the essential connection to utility and necessity as they used to. Not so long ago their use was totally ubiquitous in many everyday activities. As packaging, as laundry baskets, wastepaper baskets, protectors for ceramic containers, coffins, carriages for the wounded. Their strength, light weight and flexibility made them indispensable in a pre-plastic and pre-cardboard era.

Some baskets have never been replaced by modern alternatives. The one I remember most is one I saw drop out of the sky at terminal velocity thirty-five years ago. It was attached to a deflated hot air balloon and filled with six passengers, a pilot and three gas cylinders. I was living in France in the Loire Valley, working as ground crew for a ballooning company. I was one of two crew members who had helped get the clients to the take-off and had dragged the basket and balloon from the van, before launching them off. The basket was huge, stood at chest height, and when fully loaded was carrying up to 700 kilos. The strong Willow warp was spaced every few inches, the corners and bottom reinforced with a little timber and the whole tied horizontally together by the weaving Willow weft. The top edge was woven strongly off, and finished with a leather collar to create a comfortable and non-abrasive armrest. The selvedge was the form and strength of the basket, the weaver's craft essential to withstand the forces it would be subject to.

It was always exciting, putting the balloon up, dragging the envelope from the back of the van, unhitching the trailer and pulling the basket off it, laying it down and attaching the envelope. Then opening that up carefully and methodically, making sure nothing was snagged. Then there was the drop line to attach to the van hitch, with a complex and fun knot whose name I can't remember but which remains to this day the only knot apart from a granny that I can tie. We set up the fan and began the job of inflating the balloon, each pulling at its sides so that when Michel, the pilot, lit up the burners, there was no risk of catching the edge of the fabric.

Slowly and magically the warming air rose and took the envelope with it, a ball of contained heat rising, leaving the sagging material below it. Then with a little time, the whole lot pulled tight around the air, the basket rose onto its base and strained at its leash. Michel climbed in, his handlebar moustache and champagne persona drawing the six clients quickly in after him. At his discreet nod we slowly released the knot and, letting the rope pass between our hands, allowed the basket to rise into the sky as Michel pulled down on the burners, sending hot air racing into the straining balloon.

Thirty minutes later after a zigzag chase along and across the Loire, we stood at the edge of a sunflower field watching violent thermals pummel the balloon out of the sky. Its top pulled off, the air emptied from it; it still rose. A moment's pause and then as we ran through the head-high flowers, it dropped suddenly, crashing to the earth ahead of us, screams and moans behind the flaying sunflower wall. There were broken bones and insurance claims, but the basket was undamaged, the perishable Willow husk withstanding the fall, the woven stems sliding against one another to absorb the impact.

I didn't think of this story when I was with Hilary, her enthusiasm for her craft and its future focusing my attention. She was speaking of the ubiquitous nature of basketmaking and how in the 18th and 19th centuries it went from a village craft in Europe to a semi-industrialised craft, still done by hand but in workshops and factories, where the young were taught through apprenticeship to the experienced weavers. It also went to some degree from being the domain of women to one of men, from local needs to more global needs. The baskets played a vital role in the safe shipping of goods across the world, as well as in the making of all manner of other storage. At some stage, most goods entered a basket, be it coal or bread, laundry or vegetables. Even the cheap baskets coming from East Asia today are still made by hand, still crafted by individuals repeating designs to time constraints and often in harsh conditions.

Many of the basketmakers were blind, deaf or otherwise physically impaired. It was a craft that could be done by those who had been compromised in some way, where the fingers could lead the way, feel into the shape and task even if there was no sight or ability to walk. After the First World War it was used occupationally to aid the recovery of soldiers shell-shocked and unable to find meaningful employment, struggling to keep their anxiety and trauma at bay. The repetitive action of weaving the Willow strands not only helped their self-esteem but also acted therapeutically to still their minds. The selvedge was more than the rim of the basket, but the very threshold of themselves.

As basketry moved away from being a utilitarian necessity, as the system of manufacture and apprenticeship declined, basketmaking returned to its place as a domestic craft. In the UK, there had never been any schools to teach it, and in consequence it became blighted by the perception society had of it, as a hobby craft, which was not taken seriously. Hilary was passionate as she spoke of this deprecation, and the effort she puts into championing basketry illustrates that she is more than an individual maker, rather an advocate for the craft as a whole. She spoke of the derogatory term 'basket case' used after the First World War for the wounded soldiers who had lost their limbs in combat and were confined to basket wheelchairs. Basketry was not taught, no longer necessary, and had become disconnected from its past importance in all aspects of society.

———

Willow to be woven into baskets grows quickly and prolifically around us. The process through which these materials are converted is relatively simple, and moreover we can relate them to their source. Old baskets rot back into the earth, they are by their nature 'egoless'. Plastic baskets or containers hardly have the same light footprint, refusing decomposition yet becoming obsolete in their use even more rapidly due to single-use narratives, changing fashions or poor manufacture.

I thank my father's family for the permission they have unwittingly given me to thoroughly immerse myself in the metaphors of their trade. That we can weave fabric, 'material' as we so often call it, from the barely two-dimensional form of a thread, is a miraculous cooperation between human ingenuity and the resources of our natural environment. This basic process of weaving gives us clothing, bedding, baskets, fishing nets, hammocks and fencing, and is a reminder of the simplicity and alchemic beauty of that relationship. Warmth, food and comfort are all facilitated through that cooperation.

The thread, spun from fibres gathered from sheep, yak, cotton, nettle, hemp and Willow links us directly to the natural world through the products we use. The warp or structure of what is woven, whether stretched between trees, fastened along a loom or stuck into the base of an emerging basket enables the weaving of a story. A story told through colour and texture reliant on the weft, yet ever more forming its own language and identity, often totally covering the structure that has allowed it to come into being. The baskets and fabric of many indigenous peoples tell this story of the link between community, material and making.

When my actions have the consequence of separating a tree from its place of growth and origin, I have a responsibility to leave a trace of its ancestry, its story, even while I take the material from it, reprocess it to my purpose and leave it to the appreciation and use of others. I may do this through story, through the choice of wood in a piece, through decisions regarding the amount of reworking or through the community relationships that underlie the process of transformation.

In my own travels as a young man, and particularly those in Turkey, Morocco and Tunisia, I found myself enraptured by the poetic patterns and colours within the handwoven carpets of the tribes and civilisations of those areas. I returned to the UK from a three-month bike trip around eastern Europe and Turkey, my bike laden heavily with camel kilims, woven tribal bags that would have been used to transport grain, bedding and possessions from one place to another.

These were not simple hessian sacks, but beautiful objects woven in naturally dyed wool yarn with patterns and symbols dating back to a more visual tradition of story and belonging. To belong, *we go along with,* or *properly relate to,* a group, place or identity. When we *long* for connection, it is because we wish to *go* towards that relationship or connection. Making carries the stories of longing and belonging. Beauty, form, function, story and heritage were all woven into the cloth, so to speak. These were not objects created from a one-dimensional need, either practical, economic, narrative or aesthetic. What they embodied defines us as human.

We have had a rug hanging on a wall at our home for the last twenty years. We gathered it while on a trip in Tunisia thirty years ago. We'd gone to see the village where my wife had lived as a child, where to some degree she was searching for a sense of her own belonging. Neither of us could resist the allure of textile shopping and we spent hours in the souk in Tunis, haggling with vendors over sweet mint tea. It is a pastime I have always enjoyed, a pastime that cannot be rushed. It is a game of cat and mouse played in slow motion as the exorbitant price initially asked for slowly gets renegotiated each time another tea is served. I cannot remember how long this negotiation took, but on my trip in Turkey, I spent a whole day negotiating the price of a camel kilim only to discover I had left my money a 20-kilometre uphill bike ride away. On returning the next day to pay, the vendor feigned any memory of the transaction and I had to begin the whole process all over.

Until a year ago the Tunisian rug delighted us and our visitors with its free design, incorporating a rich symbolism where each shape had a meaning and significance handed down from weaver to weaver in the tribes of that region. It and its purchase also helped to connect my wife to her past life. It was highly irregular, created on basic looms and the variation in colour, tone and shade speaks of the hand and the natural method of the dyeing. It is rich in story in a way that is absent from all industrially woven rugs and fabric. It speaks of the symbiotic relationship between the natural world, the human hand, the tools, the maker and the community in which these objects were produced and used.

It is no longer on the wall. The natural world had considered it a squandering of resources and colonised it with an army of moths. Hidden behind its artistry, the moths slowly devoured the structure that had allowed the skills of long-dead artisans to be displayed on our wall. It was another reminder that human or non-human death is never far away, and that though humans can make, nature will always be there to repossess. The materials are only on loan before being reclaimed by insects, bacteria and other representatives of the natural world.

Chapter Four

The Edge of Wild

R ecently I was in the woods with Mike, a forester friend of mine. He is a man of the edge lands, the spaces where wild things lurk on the edges of fields and tracks, roads and forests. He is of the sun-dappled wood, where shade is not too deep to block out all light. A mass of man, with a slight stoop for all the years of tools and chainsaws, of great logs and the dragging of colossi. We are all of the light and the shade, the sparkling white of conformity and ragged grey of the dark, dank and chaotic. We are of abundant growth and the stench of death, of bright lavender sharp and the dull groan of wet leaf. No amount of education, of whatever type, can totally silence our dark feral side. It will out, and here on the margins with Mike, I feel it.

We are in a great woods, 200 or more acres, walking on a track close to the edges of cultivated fields. Metres away is the uniform yellow green of fresh crop growth, destined for markets, replete with value. In the woods, the dark falls quickly, only the dappled light at the edges where we stand, and my feral self soon itches into life. This is not the wild, far from it, as Mike and I are here to discuss the forestry practices he employs in the management of these woods. Yet, for the repressed wild in me, this is heaven, this multilayered world teeming with life. There are words we use now – ecosystem, biodiversity, wilding, reforestation – which buzz around the edges of what we have lost but still long for. Yet they are possessed by needs, by the will of systems and by a certain autocracy that means well but doesn't speak to my soul.

I feel a kinship with Mike – both of us were sent away young to be moulded to the needs of a system; both of us gravitated to the trees, to their wood, to the power of a tool in our hands, and the autonomy these relationships bring. Both of us have asked questions about how we integrate our feral selves with the needs for community and civilised behaviour.

We walk in the woods, a plan we have been harbouring since Mike became head forester on the ancient Dartington estate a few years back. I walk with him as friend, woodworker and story collector, wanting to see these woods through his eyes and passion, wanting to understand better what it takes to help 'curate' nature. I am being a little provocative, as I'm not sure that nature can be curated, but it's an idea.

I tend to prefer the notion of husbandry, the idea of careful or thrifty management; frugality, thrift or conservation. A husband was not originally a married man, but a manager, particularly a prudent or frugal one. Mike has a young man in his team called Oscar, whose job, as well as managing the charcoal burning, is to husband the young saplings that have been introduced into clearings where older trees have fallen or been felled. He will occasionally walk amongst them, register their health, cut back growth that may dominate them and sponsor them into a healthy start to their lives. He will manage them as is needed for their growth, and in a way that doesn't create too much work for himself.

I like this, the care given to these youngsters, a care we would not question for our own young, but that we might not appreciate for young trees. Within the idea of husbandry, I pull out the words 'care' and 'conservation', for together they allude to process and result, to what is put in and what is created. Implicit in this is the idea of relationship, with which Mike and his team are constantly engaged in by managing the woods in a way that avoids clear fells and the dominance of financially motivated systems, and looks more at the harmony of the woods as a whole.

In speaking of biodiversity, I understand him to mean not only the conventional meaning of it – that well-managed woods will create more diversity of species, no one too dominant at the expense of another – but also the production side of the woods, and their enjoyment by the people who access them. The word 'forestry', can easily bring to mind the idea of clear fells and a growth-and-culling policy, of ravaged areas of hillside, fresh-cut stumps brutally projecting out of compacted soil. I have often heard people speak of the 'brutality' of forestry, of the pain that the cutting down of a tree causes for many of us. I understand this personally, feel it to my core, yet there is such a danger of being anthropocentric, of placing all manner of human needs and projections in the way of a balanced vision.

As we started down the track, a thin line of Beech woods to our left tracing along the edge of path and pasture, Mike spoke of this edge, the margin between. Where pasture dominates and leaves little space for wildlife, for diversity, it is the boundaries between fields, the hedgerows, where much wildlife can shelter. Some hedgerows are narrow, with little space for the wild to inhabit. But where Oaks and other trees have been allowed to grow, where the great cutting arm of the tractor has not touched, thin woodland strips develop and with time can thicken.

The forest track we were on allowed a corridor of light to cut through the density of growth, and along its edge Mike had thinned Hemlock and Larch, the odd Cedar and Douglas Fir, to allow a little more light so that the forest floor could thrive: permitting lower vegetation to grow, Ash trees to propagate and insects to find food. With the insects would come the birds, now with lower growth on which to perch, and a breath of air would fill the woods' lungs, the dominance of darkness compromised. I would be a happy bird, given a little more light in the woodland shelter, invited to a great summer banquet of insects and grubs.

As we walk slowly along the path, caught in conversation, Mike points to a low earth bank running alongside us, set back a metre or two from the path. These woods are marked on an estate map of the 17th century, and medieval earth banks mark them further as hunting territory, a chase where the banks, palings pointing in, had been used to pen the deer. Unable to escape, they became ready prey for the regal hunting games, food for the nobility and out of bounds to all others. The trees were exclusively deciduous, fallen autumn leaves losing the chlorophyll colouring to winter's onset, fading into the brown of the forest floor. The deer picked at all the lower growth, the leaves forming a more natural feeding ground than the manicured pastures on which their domesticated relatives grazed. Estate needs left the woodlands lightly managed, trees being felled and processed for construction, saplings for palings and fences. Perhaps the First World War accelerated the felling and had made room for the plantings that have now come to dominate the visual identity of these woods.

Time has a habit of fooling us, for we meet the changes that have happened before us as if they never were, as if the state we find is perpetual. We see a place, familiarise ourselves with it and freeze it in our consciousness as if it had no past, only the moment when we were first acquainted with it. We see only our version of a story. But these woods have belonged to others – humans, wild nature and the multiple animal species that shelter, feed, play and hide in their shaded safety. Here, as in most woods across the UK, an interplay between different interests has long been a constant. When Leonard and Dorothy Elmhirst, a visionary couple, bought the estate back in 1925, they began a programme of forestry set within the context of the massive trauma of the preceding war, and in doing so changed the nature of these woods. Walking in them today is as if what seeded the change were still alive.

The stand of wind-breaking Beech trees flanking our path to the north begins to fade into the widening expanse of woodland ahead. As we walk, mature trees fading into the depths of forest dark, live green growth alongside us, I look up into the far distant crowns of trees unknown to these lands a hundred years ago. Experimental in their forestry practices, keen to quickly regenerate timber stocks, there are Giant Redwoods, Douglas Fir, Western Red Cedar and Hemlock, all American importees proliferating across the forest floor. At times they dominate, any one of them in a dense mass, and at others they interweave amongst deciduous growth. Great unbranched trunks, red-tinted, furry and statuesque, soar into the darkened space above.

I am in awe of them, these *Sequoiadendrons*, Giant Redwoods, trees that can live for a thousand years. They were brought over to Europe from the American Northwest only in the late 19th century along with all the other botanical discoveries that opened our European eyes to greater possibilities than we had ever imagined. If I stand in the lee of my imagination, I would see those giant, ancient Redwoods towering above me in their native habitat, their girth wrapping around my perception and blowing what I had believed into a new form. I would wonder how they could be on our land, in our forests, and the timber exploiter in me would wish to split them open, feel along their massive straight-grained growth and plan what I might make from them.

Here, in these newer woods, their descendants stand before me, spaced to allow some Douglas Fir and Western

Red Cedar amongst them, trees whose ancestry they share. Mike is speaking about Leonard Elmhirst and his interest in experimentation. The trees here are illustrative of the thinking that transformed the estate he and his wife bought nearly a hundred years ago. As the trees have grown to their great height, the school, crafts ventures and college that they started have come and gone. The buildings fall into disrepair and the dreams that once were are now composting on the forest floor, waiting to burst into new growth. These trees will remain reminders of those dreams for a long time, whatever happens with the man-made structures here, whatever ideas humans have for the land around them and the attempts to res-urrect what was. People will speak of non-native species as invaders, to many a reminder of the prolific spread of conifer plantations across the UK after the Second World War, yet these stands interspersed amongst native growth are a real mirror for the story we find ourselves in. The Beech stand has faded; Oaks, Ash and Chestnut whip up on the margins, amidst these non-natives, spurred into a competitive race for light, managing to coexist, helped along by selective felling.

Mike points excitedly at a butterfly drifting across our path, one I would barely notice, yet in his excitement I come to see it and what it means to him. I see a woods, I see the trees, and most I know, but for Mike this is one of hundreds of walks he has taken here. He sees each tree, their condition, their needs wrapped in with his vision. His rough-textured face is bright with the life that flutters from the decisions he has made. The butterfly in the light, on the low growth, flitting from Ash to fern is there because of the trees that have been cut down, to make room for others to grow and to let more biodiversity develop. This is not just a butterfly, but a symbol of success, a symbol of relationship. If a clear fell represents an unpleasant divorce, the promise of new relationship a mass replanting, then what Mike practises here upon the legacy of what he has inherited is all the intimacy of a relationship at work. The disappointments, the excitements and the drudgery, but the knowledge that a vision can survive and flourish through attention and commitment.

To our right a clearing of light cut through the Cedar stands, the odd toppled bough a reminder of the delicacy of decisions made, or of happenstance. A couple of trees brought down because of disease allowed winds to strike the crowns of others once protected, and because they had grown quick and whippy, unprepared for exposure, many had toppled, leaving a great hole. Now, the Ash have arrived, bird- or wind-borne, happy to grow in restricted light, and Mike's guys have planted in with a few whips. Given a few years there will be diversity amidst the barren floor of conifer dominance. The butterflies will venture in, the birds and woodland creatures reclaim the verdant cover.

The weather has been bone-dry for a while, the clay underfoot as we track off the path dry, yet deeply furrowed by tractor wheels as a reminder of the wet state these woods exhibit for much of the year. Feet dragging balls of clay up into great clumped masses on the soles of shoes, legs laden with the weight of progress, little topsoil under the conifers ahead. We walk back to the main track, compacted with shale and firm all year. Forestry tracks cut through woodland and incise it with the mark of the industry that cohabits with the interests of others. The sounds of it drift up to us as we head slowly down.

In anticipation of where he is leading me, Mike tells me of how his old woodland yard is being redeveloped for timber-related businesses, and as we come to a series of iron barricades, he shows me where he will soon move his operations. Two diggers move great piles of reclaimed aggregate, shifting it around on the scraped back floor of what was there. We are on the edge of field and forest, the space in between. It looks desolate now, ravaged and compacted, soon to be filled with the fruits of forest labour. Soon there will be piles of mature Ash butts, smaller rounds of thinned Chestnut, Larch, Douglas Fir and Cedar. Amongst them, lesser growth for firewood, to be cut and split for drying, and then delivered locally in a year or so.

Piles of butts, lying as corpses on compacted soil, each with a story to its past and future. At the start of our walk we parked in the existing yard, and walked through the piles of logs, their end grain the only voice to their identity. Rounds upon rounds piled high familiarly. The Ash pile was big, the butts small, an indication of why they were here. The chalara fungus has ravaged woodland throughout Europe, targeting the Ash and related species. They remove the affected young trees here, leaving the more mature for the cover and habitat they provide unless they represent a danger to passers-by. The young growth glimpsed through the mottled woodland light, squinted at, reveals the dead growth on the stem ends, the dieback of which the common name speaks.

There are some larger ones, too, their butt ends already sold off for hurling sticks, a traditional Irish game that needs the tapered bottom growth of the Ash trunk from which to make handles. A team of specialists came through the estate last year selecting from the trees Mike had marked for felling those that suited their needs, and Mike cut a metre or so off their ends. I see the sticks in my mind's eye, watch them as they emerge from the cut trunk, from the years of growth, from the witnessing of that tree to the passing of time. How many sticks come from one butt? It may be many, as I see them spiralling around the outer growth, and now the midsection, back another twenty years or so, and again and again for the life of the tree.

Some of the Cedar butts lying there, their broad ends pointing outwards, were big, their split grain highlighting massively wide growth rings. We'd seen their severed stems earlier on the forest floor, and I'd noted the heavily buttressed growth and the speed with which they had grown amongst the competition around them. Trees respond to stresses, twisting or bending with light competition, or strengthening their stems to resist wind exposure, or reinforcing around forks to cope with the alternate swaying of each limb. Trees are alive in their cells, alive in their connection to the forest floor and alive to the changing conditions around them. Without seeking definitions of sentience, it seems to me obvious that trees adapt themselves to changes, and that suggests a sentience of sorts on their part. The result in the cut stem is what I observe as they lie in these piles, their growth history reading like the page of a book. The splits in their drying rings speak of their over-fast and responsive growth, and render that wood useless for timber use. Between each year of winter growth is virtually an inch of summer growth, quickest where the buttresses have raced to develop. The rest of the butt will not be strong enough for heavy structural use but will make good cladding.

These trees are at their limit of size for Mike to process, for any wider than 18 inches and their weight would make extraction, transportation and cutting ever more demanding.

This is good news for the great *Sequoiadendrons* which are absent from these piles, their great girth safeguarding them from being cut, and allowing them to develop to their full potential over the next many hundreds of years. The Larch here in the woods have already been heavily thinned for the *Phytophthora* disease that they carry, and there are none here on these stacks. The other stalwart of forestry production lies here, the Douglas Fir, a great constructional timber and for Mike an important one to manage in relation to the forest economy. We passed many hundreds on our walk, and Mike spoke of all the pruning that the young trees were getting to ensure the quality of their timber when they are finally felled. I see houses and cabins rise from these piles, I see careful construction honouring careful and attentive husbandry of the woods. I imagine the spirit of homes built from trees grown well, timber processed appropriately, untreated with chemicals and built with care and skill. I imagine the meaning brought to those involved, when they are part of a creative relationship, leading to the collaborative construction of a home and refuge. Shelter is a human need; trees are a natural inclination of the land; here in these woods I see them sharing habitat and dreams.

Mike's own yard, itself piled with other butts from the various woodlands he manages, is a few miles from here up on the windswept ridges of Dartmoor's edge. I have watched his land grow over the years, sprout as if it were woodland, with the stables he first placed there, the hidden rooms above, the generator shed, and spreading piles of timber. Trees twisted and deformed in dense woodland cover, have become bones in the soul of his home, ribbed growths of swinging gates and the sinewed expression of making out of a creative connection with the will of life.

I have seen other buildings he and his men have built, great roofs of twisting Oak placed on earth walls, raw expression of the ground we live on, and the growth that expresses itself from it. The timber grows out of the earth walls as if it were a copse of living trees searching for the light, weaving their branches into a cooperative web of strength and support, cradling the roof, sheltering the occupants. I see the heritage of the woodland in these roofs. I see the churches and boats that were constructed as if the trees still grew, as if the craftworkers were as intimate with the woodland growth as they were with the growth of the processed material at their fingertips. Cathedrals of trees rising from the forest floor, arching up towards blue-grey skies, and dappled light, drawing up the human spirit and connecting us to the raw breath of life.

Oaks grow slowly, take their time to emerge from open woodland floors, find their roots and anchor themselves to the future. On our walk we have passed many, mostly alone amongst stands of other trees, many on the edges where they get the light they need. No propagators of the dark these trees, not like the Ash or Beech, which are able to wait in the crowded dusk for an opening through which to burst. No teenagers, opportunists these, not fast to take advantage, abundant with spreading seeds like Sycamore or Ash. They take their time, taking the energy to fill their branches with seeds only every once in a while. They saunter through life, steady in their pace, deliberate with their expression. They are the statesmen, have the long view, and to them many of us gravitate, wishing that we too could be as they are. Of the many Oaks we passed, smaller yet older than most of the trees here, Mike would not seek to cut, simply leaving them for the forest life. More generous to other species than other trees, richer in the life they support, their peppered presence in the woods symbolises the continuum of relationship, of meeting the diverse needs of a woodland home.

In them I see all the Oaks I have passed, all I have lived with and all I have used for my furniture. I see a woodland on the moors I go to often, its crowded bracken-covered floor full of abundant, close-packed, sky-reaching Oak trunks. Seeking the light they have grown as one, connected by a great mycelial web, their roots extending their consciousness through it, communicating their state to one another, protecting and safeguarding their community. Thin, they are probably between a hundred and two hundred years old, and in their presence time stands still.

There has been no management in these woods since they arrived here, the hundreds of acres along the river alive with an untouched aura. I feel the sacred here, for these trees, interconnected, also connect with all creatures, including the humans who pass through here on their way to the river's edge. They are the microclimate, create in their mass the humidity regulation of their environs, hold in their shade the autumn's and winter's moisture for the warmth of summer days. The moss, dank with the must of the woods, soft depth of green fertility, meters the health of this woodland home. Fallen trees are slow to rot, lie criss-crossed over paths, their trunks etched with the claws of scurrying creatures, the ground beneath scuffed and raw.

Mike's Oaks, off the woodland track, have none of this isolation, are more suburban, between the wild and the cultivated. They will witness the felling of the moneyed trees around them, and the untouched growth of those whom the chainsaw will never touch. They are part of the ethnically diverse human-centred world we live in. They are a fragment of a rich whole, a single player in a deeply layered story of the decisions made by humans, affecting all the inhabitants of this interconnected whole.

———

Recently I attended a conference on forestry and art that was held on the Dartington Estate to mark the 100th anniversary of the Forestry Commission. Mike was there in his role as head forester on the estate. The conference sought to bring together the voices of the established forestry community, with artists and academics to build a new vision for forestry practice moving forward into the 21st century.

Talk of trees as commodities, of art responses, and of the land they are on as amenity collided with each other under the timbered roof of the Great Hall. I found myself inspired yet also exhausted. After my walk with Mike I had felt hope for his vision of the forest as a communal space where human need for material resource, and the preservation of the embodied wild of the forest could find a form in which to coexist. There were those who saw the forest as ancient and untouchable, and felt the cutting of any tree as an act of aggression towards it, as if each tree was also ancient and untouchable. Is it not the tree, but the forest floor that is ancient? An individual tree is a three-dimensional projection of hidden earth rooting, of interconnection and communication. The life within the forest is much more than the cumulative value of the trees themselves. They will regrow, and those that fall or are felled can contribute to the life of the whole if properly managed.

Three days spent talking about trees and forestry and no one mentioned the great timbered roof of the Great Hall we were gathered in, no one mentioned the engagement with the made, and our necessary reliance on it. Amongst the forestry community and the artistic community, the made and the maker were given little voice, as if all were embarrassed to talk about making and its reliance on raw materials when the talk was of conservation, production and the dignity of trees.

When making happens 'out there', when the massive plantations on other lands invisibly supply our consumption needs, and we are disconnected from our involvement, we become illiterate in the art of living together, human and non-human. Houses are built from branded materials whose heritage is entirely inarticulate, has no language with which to communicate. People mourn the trees, wish to hug them and talk to them, speak of ancient forest and then go and shop at Ikea.

Big trees, old growths, will capture the imagination, but the lower-layered, the straggly and those fighting for light, that may be where we need to listen. Below the institutions, the financial imperatives, the gilded halls and ivory towers, who is it that works the earth, that remains connected to the graft and grime?

Within the commons, and from the idea of commons, there are stories of connection with the understory, with the metaphors and abstracted memory of subsistence eked from the forest floor. The rights of pannage, estovers, agistment, piscary and turbary are like terms pulled from the dictionary of our connection, ideas imbedded in words through which the people formerly attained some sort of living wage. People called Smith or Cartwright, Butcher or Turner. Our people once subsisted from the land and owed their names to that connection as we, too, owe ours to them.

In a time before the Latin language gave us names to which we now have faint attachment, our people had the right to roam, the right to collect firewood and fish, to graze their cow or two so they could get their milk in exchange, to put their pigs out to gorge on acorns or to hunt for rabbits or pigeons. Norman kings took, Henry III gave back, and since then all has been taken again under the banner of progress. Somewhere here in the understory of past consciousness, under the layers of dirt mulch, there are seeds just waiting to burst into life, to remind us of the power we have, the connection we had, and what is never lost.

I imagine those who once foraged for wood, who turned their dedication to the managing of small sections of woodland, to coppice practice and small woodland crafts. Objects made for the home, bodged chairs made on a pole lathe and shaving horse, became produced by dedicated bodgers for market demand. The poetry of tree growth, the unremitting energy of nature, together with the coppicer's attention allowed for a synergy between nature and people's needs.

There was a woman at the conference, nearly forty, who had been an artist, did some care work and was enrolled in a coppicing apprenticeship. During one particular talk by various established voices from the forestry world, she asked what was their idea of coppicing and the sponsorship of the young to learn. The answer given was polite and somewhat condescending, cloaked in general ignorance and awash with an implicit judgement of the futility of such practices.

Coppicing is a wonderful example of our ancient cooperation with nature, a direct path into a human-nature relationship that is beneficial to all. A tree once cut mature for its butt, can regrow, not as a single stem, but in multiplicity, many competing with one another for strong and rapid vertical growth. The light well created by the felled tree gives them the energy they need for unremitting growth, all surging in unison. Cycles of multiple years, of felling and regrowth, extend the life of the stool to many hundreds of years and speak of an ancient relationship with the trees for the material they produce. Firewood in plenty, or components for the bodgers.

That young woman was looking for relationship, searching for connectivity, for a way to make sense and meaning of her life. She imagines that one day she will work in the woods with at-risk women, helping them build their confidence and connection through the act of care and husbandry of the coppice. She, along with others, will help change the story of why we work in cooperation with the woods.

As I reflect I see her as a coppiced stool, cut from a single growth by her own volition, to see what would regrow. The cutting and subsequent new growth are an opportunity for new influence, for energy to emanate purely from the spirit of the being, tree or human, which will erupt into a new form. I do wish I could remember this woman's name, though perhaps she has taken a new one now to match the changes that she's inviting. I will call her Naomi.

Around Naomi in that great hall where the conference was staged, were countless single-stemmed trees: statuesque, with great crowns, male and well positioned to take the light from younger growth. They were grouped together, dignified, yet in their mass not allowing young growth the light it needed to thrive. The old coppice amongst them, redundant for years, was barely alive, diminished and leggy, invisible to all who came to gape at the taller trees.

The conference's duality merely reflected our present situation, where the old and dominant perspectives of production growth and monoculture are being challenged by a growing advocacy for the planet and for all who live here to have a voice and be heard. Naomi stood fairly alone in that conference, yet I have met many versions of her, many men and women who are in effect asking questions about their own growth, whether they want it to be single-stemmed in the image of what has been. How they make their living, how well they live, and whom they touch and positively influence will come together to create a life of meaning where trees and the environment have the primary role.

We no longer live in a time when such a relationship with the woods is necessary from a practical viewpoint. Coppicing and tree management for craft purposes and firewood production are no longer imperative. Timber was not so very long ago the only fuel with which to heat homes, smelt iron ore or melt glass. Carts were built from it, as were their wheels, the spokes cut from coppiced growth.

In a time before corporately controlled power genera-
tion, timber was the material of innovation and freedom.
Growing, cutting, processing and constructing with wood
were all reasonably accessible practices, allowing degrees of
independence and practical expression. Centralised power
production, and the industrial development of metal and
plastic usage have contributed to the relative shrinkage of
the timber-based industries.

There is no other construction material like wood,
where we can still play a role in all stages of its life, from
seed to produced item. So, coppices became redundant,
overgrown stands on woodland parcels adjoining farmland
or even flattened to make room for expansion. Coppices
have become copses, dense and often unmanaged woodland growth. Copses have become corpses, overly shaded,
barely alive.

When I was growing up, I was given to understand
that the phrase 'a bodge job' meant to have done a
bad job, to have knocked something together, to have
'botched it'. I have spent the last year asking dozens of
people – my students at university, friends, acquaintances,
industry professionals, academics, delivery drivers, fellow
café dwellers and artists – what they understood 'bodge'
or 'botch' to mean. Recently I even asked my eighty-
eight-year-old mother, convinced that she would know
the answer.

The responses have generally been very conclusive. Both
words give rise to an understanding of a job poorly done,
but 'bodge' leans more to the idea that the job is adequate
for the needs, while 'botch' quite definitely suggests a

shoddy job under any circumstance. 'Botch' arose out of 'bodge', which itself comes from the noun 'bodger'. The bodger was a person, not a job badly done, a craftsman who made his living in the woods.

Right now the spellchecker is highlighting 'bodger' as if it were wrong – not that it's tried to correct it, only that it doesn't recognise it. In an algorithmic world, it does not exist. As sawing great trees into planks became more efficient and faster with machinery, and as straight-line thinking became more dominant, it is easier for producers to conceive of components cut out of planks, than those split with human energy from coppiced growth.

Do an online search, however, and you'll find some images on the great web of these maligned monotone men, faded pictures of a past life, piles of wood and ramshackle huts amongst the trees. The men, on shaving horses, leaning into a pole lathe or splitting shingles, were a perfect synergy of our making needs and nature's habit of growth. The cycles of coppicing gave them the material that they required, the divergent species creating specific product relations.

Chestnut and Ash both coppiced well, and both split easily on the bodger's block. The former could be used for gates and fences (hurdles, as they were known), the timber resistant to rot and ideal when left in or on the ground, whereas Ash would all too willingly rot. However, Chestnut was not so able to absorb impact, more fractious and liable to crack, so Ash was used for the spokes of carts, where its pliability allowed it to resist the impact on the carriage wheels. The Ash spokes were held by the central Elm hub, a wood not prone to splitting at all, and as such unsuitable for making long components. Used on cart hubs, bellows and Windsor chair seats, Elm was ideal where integrity was paramount, either over great widths or as a meeting place for converging forces.

As humans we seem to organise our experiences with narrative construction, creating stories through which we define good and bad, what serves us and what doesn't, whom we wish to admire and whom we would denigrate. We create a sense of ourselves both individually and collectively by story making. When it comes to our view of nature and our relationship to it, the prevailing post-industrial narratives have inevitably dominated, leading to acts that have tilted the balance in favour of the human and not the natural world. I wish to bring the stories of our connection to nature to the fore, to make conscious our use of its resources so that our endeavours to transform materials can be beneficial to all.

Many of us are probably more able to recognise a species when it stands alone and has been able to attain a shape and form unhindered by competition with neighbours. An Oak standing alone in a field, majestic in its hundreds of years of age, establishes itself in our minds as the quintessential tree. The tree that a child might draw, with far reaching branches, a great rounded crown and no more height than width, its trunk short before the first branches reach out. There are the Rowans up on moorland; Mountain Ashes, small for their years, berries on their gnarled boughs bright orange, their crowns sculpted by the prevailing winds on the exposed sites where they choose to grow. I think of the great estate trees, the great mature Walnuts, the Oaks, Limes, Ash and Chestnuts. There are the Yews in graveyards that trace their ancestry back to pre-Christian times, when they enjoyed sacred status amongst the pre-Roman inhabitants of this island.

There is a ruined church near where I live, its belfry intact, the rest of it repurposed in a wall or necessary and historic repair. Behind it, shrouding the graves over which it grows, is a great Yew tree, ancient in its gnarled and twisted growth, its dense, dark needles austere and solemn. It predates the church, itself eight hundred years old, by a thousand years and speaks of the spiritual connection between humans and trees prior to the rise of the Christian faith. As one of only three conifers native to these islands, the Yew had a significance for the Druidic world long before the arrivals of the Romans. It is a gateway between life and death, a portal into the world beyond us. It was also the resource for the makers of local longbows, Yew wood having the prerequisite qualities of memory and spring. This tree, magnificent in its size (23 feet in girth), is time held still, two thousand years caught in the physical presence of a bough, the material and spiritual alive alongside the human.

It has lived through a time when Druidic customs laid the foundations upon which later Christian ones were built, where owning a bow and practising archery were not merely the custom but actually the law. This tree has seen empires come and go, monarchs rise and fall, and the land it is on go from being a parcel gifted to a subject to its many incarnations since. While everything else has adapted around it, while other trees have come and gone, this one, gifted by its own genetic heritage and its meaning to humans, has remained as it ever was. It has had our protection because of the story to which it is attached, to its place in our collective consciousness as a tree of spiritual significance. Simply because of its age, it has become worthy of our respect.

I would wish that all trees were worthy of our respect, but in a worldview where they can be resource, memory or wild nature; where they can be impediments to development or symbols of its resistance, the picture is infinitely more complex. Here in Devon there are a lot of Sycamore trees. A resilient survivor and self-propagator through seed and suckering, it quickly establishes habitat. A neighbouring farmer of mine years ago introduced me to the idea that the Sycamore was locally known as 'Devon weed'. It had this reputation in the local farming community because of the speed with which it would invade untended land, and as a result of the bias that deems plants that are not invited in by humans 'weeds'. As such, it is shown no respect, cut down at will in a way that would never be acceptable for an Oak. The Sycamore represents wild nature seeking to re-enter, repossess, and it is unwelcome by those who manage the land.

The Sycamore had no absolute role in old crafts as did Elm for its resistance to splitting or Oak for its durability. Yet to a maker like myself, and to a maker of stringed instruments, it holds a hidden and subtle quality that makes it irresistible. When harvested in the autumn and cut not long after felling, its cut face is creamy white and its texture soft and smooth. Dried with care, vertically and not horizontally, it will preserve its colour and reveal its subtle beauty ever more when the artisan reworks it.

However, treated without care and reverence, dried improperly or left too long in the wet, Sycamore will become grey and worthless, a wood keen to return to compost as it would quickly on the forest floor. When a specimen has a particular grain pattern (a 'fiddle', as it is often known), makers of fiddles and stringed instruments will pay richly for the unique qualities it brings. To the rest of the industrial world Sycamores have little significance, and there's no prevailing narrative that will safeguard it from becoming anything more than firewood or mulch.

I cannot but look at the world around me from the perspective of a woodworker. I see the trees that shade me, recognise them or not for their species, yet quickly my eyes wander to their trunks, to the bottom growth of the tree from the ground up to the first major boughs. Is there enough clean growth on that Oak to make a big dining table? Is the burr growing from it tight enough or interesting enough for detail work? I may see a large field maple standing tall and dignified and examine its bough for signs of a ripple, for a variation in texture that would reveal itself in the cut planks. Their shimmer entices my interest in the tree, my imagination building a sense of the texture and sparkling beauty of the finished piece.

While my wife looks to the spring flowers pushing out of the earth, I see the gently twisting form of the Ash tree standing majestic in the woods we walk through, its mottled bark pristine on its trunk, shadows defining a gentle twist as it towers above me. I wish to share it with my wife, but am concerned that my enthusiasm will bore her, that from her perspective the blues and whites and yellows of glorious coloured new life are much more interesting than yet another tree trunk that may enthuse her husband.

Even that Yew tree in the Dartington churchyard, its great squat bow rippling and occluding as newer growth folds around older, excites the woodworker in me. I think of Yew planks I have in my workshop, their red, tight and smoothly textured grain suddenly penetrated by black lines where the occluded bark of the twisted tree bursts through the great straight-cut plank. I feel the impatience within my furniture maker's persona as it puzzles out how to cut that plank, what to extract from its tortured surface and whether the quality of the plank was worth the life of the tree.

The hobby naturalist in me wonders at the reductive thinking that would take such vibrant and natural forms and place on them the straight lines of post-industrialist thinking. The saw cuts recklessly through the years of growth and the structural integrity established through the life of the tree, growth patterns that speak of the stresses and life challenges that it has survived. The beautiful marks left on the surface of sawn planks are a portrait of this disregard, exposing fractured impressions of growth layers to the scrutiny of the viewer. Yet they remind me forever of the tree they once were, and when that tree has come to the end of its natural life, the piece that is made from it will be its representative wherever it finds a home.

Chapter Five

From Bough to Boat

A t the heart of my enquiries into the world of materials is the idea of relationship. The relationship we have with what we use, the relationship that we have with the world around us, the relationship that what we make has with the materials used to make it and the relationships that develop around the process of making. My personal focus in this entanglement of possible relationship has been with trees, and the wood we reduce them to and potentially elevate them through. Trees become wood, and wood becomes furniture, carts, ships, machines and almost anything we can imagine. Our landscape has a natural tendency to want to become woodland again, to return to what it was. The pull of natural forces lies at the edges of all fields, behind all fences and on the margins of human endeavour.

The largest unfarmed and forested area in the UK is the New Forest. I've travelled through it on countless occasions, on my way to see family members who live there. Mine is not an intimate connection, only a passing acquaintance as I move quickly through it. There is the necessity of being mindful of the itinerant beasts who wander carelessly over the land and onto the roads with no sense of a boundary between the two.

The ponies, cattle and sheep, here owing to the continuation of commoners' rights, wander freely amongst the scrub, marsh and woodland with which they share the landscape. Like Dartmoor Forest, its name is a reminder of the Royal ownership of it proclaimed by William the Conqueror in the 11th century. An ownership that continues to this day, and that despite commoners' rights in a good portion of it have dominated the use and relationship others can make of it. About 20 percent of it is postwar coniferous plantations, made to compensate for the widespread felling of deciduous trees in the early 20th century. About 40 percent is deciduous wild forest of different ages. The rest is the land over which the domesticated beasts wander. Since the 12th and 13th centuries the gross coverage of forests in the southern part of England has massively diminished, leaving this as the largest remaining section. It has continued to have a difficult relationship with its own status as forest, the needs of king and country pulling at it for material procurement.

On my most recent trip I was combining pleasure and work again. A family gathering in honour of what would have been my brother's sixty-first birthday was taking me the 150 miles across country from Devon to the quiet heathland that is the first experience of the forest, driving into it the way I do. The quiet of these last lightly touched spaces is the first thing I notice, away from the bustle of main roads, into a version of landscape that celebrates the natural world. I was also coming down to look particularly at the human interaction with this landscape over the preceding three hundred years or so. I would drive the following day down to the edge of the forest at Beaulieu, part of the 10 percent of it that is under private ownership and not that of the Crown. There I intended to visit Buckler's Hard, a location once used to build wooden warships and, for a hundred years or so, one of the most important private contractors to the navy.

Beaulieu is on the river of the same name, a snaking pattern on the map that feeds into the Solent near the Isle of Wight. Streams run into rivers, rivers into larger rivers and those into the sea. There are similarities between this system, the human circulatory system and the root system that supports the growth of trees and forests. Here is part of the alliance I seek, where the human, the arboreal and the features of landscape are all in necessary relationship with one another. Men have built ships from wood to sail in rivers and oceans for thousands of years. The act of sailing through water on a vessel is one of extraordinary participation between various systems of life. Without this cooperation we would not have set foot on distant continents, nor would we today benefit from discoveries that furthered Western understanding, knowledge and science.

The twists and turns on the Beaulieu River created particular currents and, through that, features that suited humanity's endeavours. Buckler's Hard was named after the fact that the bank at that place was firm, could be landed at and launched from. Moreover it had some shelter from the winds, a tidal pattern that suited the launching of vessels, a depth of water free of silting and near it an area of forest from which timber could be sourced. We found what we needed from nature in this place to support our will.

This relationship is so often captured in the names of places. Buckler's Hard will forever speak of the importance of a firm footing on the river's edge, as Beaulieu tells of a beautiful place, *beau lieu*. Names preserve the past, even if their resonance fades over time. They are a valuable reminder of what pulled humans and place into relationship in the first place and communicate essential and lasting truths.

So Buckler's Hard describes a firm footing on which boats could be birthed and launched, on which ramps for construction could be built, and on which business endeavours could rely. Arriving there, at first I observe little that speaks of this early relationship other than the obvious presence of the river. A series of attached cottages line two sides of the wide street which goes down towards the river's edge. From the last of these on the left, as I walk down, people spill and enter, mill and sit on benches along its face. This, the old master shipbuilder's house, is now a pub, host to the likes of me, people drawn to the preservation of this place and to the stories it holds and the 18th-century beauty caught within it. This village, once a thriving commercial centre of boatbuilding, is now primarily a tourist destination. As in the granite quarries on Dartmoor, the old preserved tin mine at Geevor or the clay works near St Austell, old methods of working relationship have been preserved here only through the very different purpose of a leisure activity. We have gone from being makers of things to being observers of how things were made, one step removed. I am glad of it, glad that anything has been preserved at all.

In the little museum at Buckler's Hard, amongst many more eye-catching exhibits, there is a small drawing that shows an Oak tree. On its branches are drawn the components of a ship that could come from those parts of the tree. The branches bear such labels as KNEES, RIBS, and KEEL, and identifies the tree's final purpose – not only ship's components but also material. The Oak tree holds a place of importance within the very core of English character; its hardness, resilience and the growth patterns of its branches allowed for a seamless integration between the ribs of the boat and the keel. The great might of the warships hung on their Oak and Elm structures, the very nature of the tree being transferred to that of the ship. Air, water and the sun gave the Oak life as a tree, the earth supported it and now we were able to take from it and the way it had naturally grown the structure needed for building. The nature of the ship's design had its roots in the 'nature' of the tree's growth. Moreover, onto this was overlaid the nature of the British character, the perceived resilience, fortitude and indomitability that allowed it to build such an empire.

The vast majority of the New Forest was owned by the Crown, and that meant that the wood taken from it in the 18th and 19th centuries was used solely by the Crown boatyards. Private contractors like Henry Adams operating at Buckler's Hard would not have had access to thousands of acres on his doorstep, and the wood he did have access to on the Beaulieu Estate of the Montagu family was not sufficient for the needs of a busy boatyard. It took a staggering two thousand trees, or 40 acres of woodland, to produce the material for a sixty-four-gun naval warship. It is extraordinary to take this information in, to imagine walking through 40 acres of deciduous woodland, to imagine how long it would take to cross, how long it had taken to grow, and the life in and around the many trees in it. Then imagine a ship only 120 feet long or so and 30 feet wide and how those 40 acres may fit into her!

This is just the raw material and doesn't even begin to address how the trees become wood, how they become ribs or knees, planking or braces, the labour that would then have gone into the conversion of them, or the energy it took. This process started with the selection of the trees, a job done by the timber trader. In the case of Buckler's Hard this was Adams, the shipbuilder whose other trade was that of timber merchant. Contracts earned with the navy were gained only through the assurance that the shipbuilder had enough quantity of dry and quality wood to ensure that there would be no delay to the building process. As the timber needed several years of air-drying it was essential that they held plentiful stocks of it. So they were always working ahead of themselves, procuring suitable trees, shipping them, planking them and drying them. This was much the same for many trades, including the carriage builder, wheelwright and village carpenter, but in the shipbuilding trade, the sheer size and quantity of timber needed were on another scale altogether. Henry Adams in his role as timber purveyor would have hunted around the South of England, and sometimes further afield to look for suitable trees, at a time when, because of wars with France and Spain, there was great competition from others over the available wood.

Ships created only a portion of the overall demand for timber. At that time, before the widespread use of coal, timber was heavily in demand from the domestic, glass, lead, iron and brewing sectors of the marketplace as a heating method. It was also used for constructing houses, bridges, docks, military fortifications, farm implements, gates and furniture. The consequence of this widespread use of timber before the advent of coal and iron was that ever since the Middle Ages, tree stocks were diminishing, and by the 17th century, men such as John Evelyn in his work *Sylva* were calling for an urgent and organised system of new planting, so that future needs could be safeguarded.

As Henry Adams went on his hunt for the wood he required, he would have held within his mind a picture of all the various sections of the boat he would need to supply, and the trees that were suited to them. He would have looked at their girth, at the growth of their major limbs and at any defects or signs of fungal attack. Each tree represented a huge investment in money and time, so a careful choice was vital. Risk was ever present, yet with skill and time the best choice was made.

Then it was for the cutters to come and fell the trees, taking off the major and minor limbs, sawing them carefully to the specification provided. The sites the trees came from would always have had to be considered carefully ahead of time, because this was the age of horse power and wind power alone. No engines here to drive great trucks down highways, or tractors that would effortlessly pick up the great logs as if they were matchsticks. All was done by human hand with the help of the horse alone. It would have been done as it had always been done for the thousands of years preceding. Horses would have pulled the trees from the woods to a loading area. Then it was down to men with levers, aided by gravity if possible, by steers or horses if not. The lengths of timber would be skilfully loaded onto carts, care being taken to ensure that the cart did not topple, and no one was hurt.

Just 40 cubic feet, or a ton weight, could be hauled by a cart, and for every ton of naval shipping, one and a half or two loads of wood were needed. So, the HMS *Agamemnon*, the most famous ship built at Buckler's Hard, would have needed two thousand loads of timber! This harks back to those 40 acres of forest. Though the biggest ship built at Buckler's Hard, it was by no means the biggest naval ship of the time.

As a woodworker I am in awe of the magnitude of this undertaking, knowing how much manhandling the timber from just one tree takes, how much interaction is required between the delivery, sticking, drying, moving and using. That a ship like the *Agamemnon* would have taken two thousand or more trees – that these were all cut by hand in saw pits, sticked and stacked along the main street between the houses – is astonishing. Each year a boat of considerable size would be built, and alongside the timber for it was all the other timber drying for future projects. Five ships could be built at the yard at any one time, on the five slipways whose remnants are now barely visible.

Two thousand journeys were needed to get the timber directly to the yard or to a waterway by which all the loads could be shipped down rivers or canals to the sea, and from there to the Beaulieu River where the boatyard stood. These journeys by cart could not be long, a few miles at the most – otherwise, the economics of transportation would be unjustifiable. These men, from Henry Adams through the sawyers, carters and all the various grades of framers and shipwrights, were in an everyday collaboration of bodies, minds and material towards a final task. Together, with all their different skills, their formidable team was capable, under the master shipwright, of completing what from a modern perspective was the extraordinary task of building such a huge and heavy wooden vessel.

Once the timber arrived it was unloaded into ungainly piles of straight and twisted stems. Some of these logs were straight, but many were shaped, as they were intended for the curved parts of the boat's skeleton. These pieces needed to be sawn to width before they could be laid aside to dry. Before this happened they needed to have been lying in butt form for a year or so to allow the sap to sink and to begin the process of acclimatisation. The yard would have had several saw pits, some higher up the slope to do the initial processing, others nearer the slipways where final dimensioning would have happened. The trees moved from the woods to the yard, and progressed through the yard as they became ever more processed into the exact material and then components required for the task.

Seen from above on the wing of a bird, this anatomy of the yard and of the journey the trees took would be clear.

Piles of unprocessed butts at the upper end, sticked and stacked piles of planks in the middle, and then more piles of useable components closer to the ships under construction, and then the ships themselves, skeletons at various levels of formation rising out of the ground. Around all of this, men would have bustled, busy with their tasks, supervised by the foreman, overseen by the overseer sent by the navy and the whole coordinated by the shipbuilder himself.

Amongst those dead trees stacked throughout the yard would have been various species. Many of them would have been Oaks, some up to 3 feet across, many curved, some including a branch forking off the main stem. These are the compass timbers that would form the skeleton of the ships onto which the cladding and internal structure were attached. Oak was chosen for the forms in which it grew as well as for its strength and resilience and resistance to rot. Then there were the Elms, ideally great 25-foot-long butts, up to 3 feet or more wide. These would form the keel of the ship, scarfed together to form the total length required. Elm absorbed impact well, did not split, and grew long and straight. It was ideal for the keel. Then there was Beech or possibly Ash, which would have been used for infill on the inside of the ship, lining over the ribs and strengthening the hull. Neither Ash nor Beech are noted for their rot resistance, but away from direct water contact and without a primary structural role, either would have been an appropriate choice. Whether or not wood will tend to rot depends on the rate it can absorb water. Absorption is affected by the density of the wood, by the presence of oils or resins in it or by its tannin content.

The woods most effective against rot are those that grow in the tropics and are both dense and have a high oil content. Woods such as Teak, Iroko, Jarrah, first imported into Britain in the 19th century or so, when used in the boatbuilding industry, created very resilient vessels. Vessels like the HMS *Trincomalee,* still floating today more than two hundred years after its construction. Built in India from Teak owing to timber shortages after the Napoleonic Wars, she is the oldest British warship still afloat.

At Buckler's Hard, Oak was the most resilient wood Britain could offer. It was protected from rot and insect infestation not through oils but through its high tannin content. For the mast and cladding, softwoods imported from the Baltic, close-grained and huge, were the material of choice, as they were more fit for the purpose than the deciduous trees then dominant in Southern England. The trees selected for the masts would have been hugely long, massive in girth and very expensive to purchase, ship and work.

I am trying to give a sense of the ecology of the boatyard, of its nature, derived from the wilder nature beyond it. The yard was a sanctuary within the wildness of the forest, a place of endeavour and transformation, where a tradesman's bodily and mental facility allowed for the conversion of materials to purpose. The floor of the yard, alive with various stages of timber in transition, was a little like the forest floor, on which decaying timber and vegetable matter were broken down by moisture, time and fungi to form the soil on which new life would grow. In the yard at Buckler's Hard standing timber was manipulated to take the form of a ship, generating jobs, industry and development – and all their positive and negative consequences.

There are remnants of the slipways on which the boats were built two hundred years ago, but only one is relatively intact. A model in the museum shows how the village would have looked when it was a boatbuilding centre, the main street full of wood, the houses dwarfed by the ships in construction. Around the village there were great piles of logs, animals grazing and vegetable plots. It was a community of shipwrights, carpenters, sawyers, carters, carvers, overseers and general hands. The ships that rise up out of the model rose to full size on the now decayed slipways a long time ago, and the workmen whose model figures hit, saw, adze and carve really did those things day in, day out, each acclimatised to his role like a cog in a smoothly operating watch. Work set the daily rhythm of village life; time ran to it as well as it to time. The hammer blows started at dawn and finished at dusk. Rain and sun accompanied the tasks, little shelter from either available to the men working to naval deadlines.

I might imagine myself in this scene, a model come to life. What role would I choose, whom would I want to work alongside? I would choose to be an apprentice to the shipwright who is setting out the knees onto the Elm keel that lies massively on chocks along the sunken slipway. I would be set up near the skeleton that was slowly coming alive. My instructions were to dress the ribs ready for their fitting to the knees. These forms would have been mapped out from the drawings, and copied out onto patterns in the mould loft, where there was ample dry space. I would have selected some pre-cut shapes from a pile, inspected them for soundness and the appropriateness of their form. I would transfer the markings from the template to these forms and then get my tools prepared.

I assume that the forms were already close and I am merely adjusting them with the help of the adze. I would have my own canvas bag of tools, made of robust hemp. I would draw out of my bag a couple of adzes, one long-handled, with a wide flat blade set 90 degrees to the handle, which protruded through the robust back end of it. Another was smaller, with a narrower curved blade where more material needed removing. Other tools lying in the bag would have included a broad chisel, a gouge, a set of compasses and a mallet. It was not the tool bag of a furniture maker, not tools of great sophistication, rather tools made by the apprentice with the help of a local smith.

Selecting the large flat adze, I would position myself over the curved Oak form lying on the ground, and standing with one foot on it would swing the adze rhythmically as I worked from the back of the piece to the front, each blow of the tool removing a clean chip that ended in the hollow left by the previous blow. As I moved along the rib, it would gradually narrow and conform more closely to the pattern whose shape dictated its own. All control was in my relationship with the tool, with the confidence in my swing, the relaxed and intentioned movement that I would initiate and that gravity would aid.

My nature would ideally be in harmony with that of the tool, the wood and environs I worked in, the air and light of the day. I know from my own work that a tool like an adze, without guides or sophisticated engineering, soon reflects back to you if there is any doubt, any tiredness, any lack of proper preparation in the wielder's endeavours. If my body were not in harmony with the environs it found itself in, with its position over the wood, with a thinking mind to hamper it, I would struggle to work smoothly. The adze, hitting the wood just at a slightly incorrect angle, would either bounce off or dig in too much. For each blow improperly done, my body would tense and compound the mistake. All this, chip after chip, to do a job that today is done on a bandsaw or computerised milling machine in a fraction of the time. Imagine the skill it takes, though, the risk and the need for total presence. Hard labour it would be, yet within it, to really make it work, there is something that transcends labour, a spirit which connects the human, the task and the transformation. There is magic here.

The ship grows from the individual components and efforts of each craftsman involved in the process. Over a year or so, a large ship – already years in the making in terms of planning, financing, and the procurement and seasoning of the timber – would gradually grow out of the keel on which it is first laid. It is the backbone of the craft, and like the backbone of an animal carries the weight and purpose of what attaches to it. The strength and integrity of the ship rely, as they do in the animal, on the skeletal structure within which its purpose lives. The skeleton carries organs, muscles and ligaments; covered in skin, it is smooth to the touch as it wields influence over a lifetime. So, too, the wooden vessel, crafted from wood, skeletal in form, holding and housing objects, products and the humans that work in it, the whole covered to be watertight and smooth, to cut through the waves and fulfil the purpose for which it was created. When finished, the chocks holding it are hit out of the way on a high tide, and the unmasted boat is floated out down the river, first to be masted and then to set sail on the open sea.

Chapter Six

Fluid Earth

On an overcast summer's day last year, I found myself squatting on damp ground in Mid Devon holding a large pot in my hands, turning it, feeling the weight and tactility of it, its texture and character. Stone smooth in places, as if worn by thousands of caresses, and in others rough to the touch, particles imbedded in it, ash-glazed and raw. As if it itself were a rock, exposed to weathering on the one side, grown from coarse earth on the other. I weighed it, and I was moved by the weight I bore; by the craftsman's touch, by the immediacy of a movement, the vision of an outcome, the dashed dreams and restructured hopes as it goes with him on the path to its creation. As I turned this pot again, over and upside down, aside of the scallop shapes pressed into it, there was a scrawl, an identity scratched in the hardened clay. This spoke of the maker and reminded me of why I was here.

I have always loved the feel of clay vessels, especially those through which the hand is felt. I've bought various pots, cups and jugs over the years from many places around the world. Of all the incognito potters who have contributed to my interest, there is one name scratched into several jugs and mugs hanging in my kitchen: NIC COLLINS. I had never met Nic till recently, but the form of his signature is clear in my mind, has been for many years. I picture it and recall the pots that I have of his, and the resonance of the experience. I almost envy him, for the freedom, unpredictability and earthly quality of his work. I envy him because to some degree I yearn for that wild soul to enter the work I do, to fracture the control and perfection that can blight the furniture maker's craft and disengage it from the trees it grows from. It was inevitable that I would find my way to Nic Collins' pottery one day, and equally inevitable that he would feature in this book.

This is why I found myself turning over the weighty pot amidst the weeds in Nic and Sabine's garden. They live amongst their pots and pottery. After heading up the bumpy drive and parking behind his Volvo, I walked out into a world of pots. Pots on benches, pots on the ground, pots growing amongst the weeds and pots peeking in their raw state out of the pottery. I saw the edges of kilns, felt the resonating heat as I passed by one on my way to the front door of the house, and one quite cold that was in the middle of being loaded.

Before my visit I had searched for him on the internet, knowing very little about Nic outside my own interest. Amongst the images that surfaced around his name were many of a handsome young man on a drum set, posing for the camera on his own or with an older bearded man. Peppered amongst these were images of pots and jugs, plates and great platters, bright with earth pigmentation and striated through with a signature of heat. The young musician was the son of Phil Collins, of Genesis fame. Scrolling down I found another grey-bearded face, etched with an intensity burning into the camera lens. I found the potter in the pots, and the pots in the potter, as if his face were another object within the catalogue of his wares.

I knocked on the door and was greeted by Lucy, Nic's apprentice, who went off in search of him. A brief glimpse and a hand waving to say he would be along soon, and I scuttled amongst the kilns and scattered pots, getting a feel for the environs of his space. Stacked clay in open sheds, the cob-built pottery, three kilns, polytunnels and a camping field. The space reeked of making, was texturally rich with the life that it took to make – to make the pots, to store, to fire, to live. I find this again and again with the makers I visit, many who work from their homes, for whom the making is as much a lifestyle choice as it is a craft and identity.

When Nic appeared and we had greeted each other, I explained why I was there and we settled on benches in the garden with cups of tea. I asked him why he makes. It was a simple question, almost banal, yet possibly complex. He smiled, as did Lucy, who was sitting with us. 'It's a lifestyle choice,' he says. I question him, and he speaks of being unbound of expectations, free to sit in the garden talking with me over a cup of tea, free to make his own agenda, free to move amongst the different tasks that define his practice. The relationship with the wood, with the wheel, the clay, the kilns, with apprentices, assistants and students. With sleepless nights of kiln-watching, the pleasure of nurturing the heat of the fire, the excitement and fear of opening the kiln after a firing.

Behind us, near where the cars are parked, was his woodpile, huge, neatly bundled and stacked, drying for later use or already ripe for imminent firings. The kiln that I'd felt the warmth from was cooling after having been fired a few days before, still sealed with three months of his work stacked carefully inside it. I never did ask how much wood was needed to heat it, but I do know that Sabine's much smaller kiln will take thirty or more hours of firing to reach temperature when she gets it going in a few days. That's for a quick fire, so I can only imagine that Nic's takes much longer, and in doing so consumes a huge amount of wood, a few cubic metres.

As we chatted it became clear that the wood was the centre of his practice, that the fire and heat it created were the real 'material' of his craft. Not to diminish the clay or his skill, but as he talked, I saw that he was of the heat and flame, lived for them, so to speak, and for the unpredictable quality and texture that they imbued his pots with.

This was where his creativity and passion lay, within the wood-fired aspect of his work. So when a sentence or two later he told me that he'd heard only the day before that the source of his timber had dried up, I felt the implications. Wood firing without wood would be much like my work without wood. For years Nic had sourced his timber from slab wood left over after sawmills process logs into planks and beams.

It would have been mostly Larch or Douglas Fir, and some years ago it was valueless. Yet with the fears and challenges of climate change, discussed in phrases such as 'carbon neutral', 'sustainable' and 'biomass', have come constant price rises, and now increasing scarcity as massive biomass electricity plants, converted from old coal ones, sweep up all the homegrown softwood supplies. The sawmill local to us – which had sawn all the wood from which we built our workshops, house, studios and cabins - is now struggling to get hold of logs at all. So we go on heating our houses, generating electricity, supposing that it is being done in an ever more ecological manner. However, carbon is still being pumped into the atmosphere, and our overground resources are increasingly being stripped while overall consumption goes uncurbed.

I'm not sure whether it's entirely fair to compare his reliance on wood to mine. Yes, we both rely entirely on it, and it allows our work to be what it is. For Nic, it is the effect of the wood firing that gives his work the complex texture and somewhat unpredictable outcome that distinguish it. Rather than simply heating his kiln with wood, Nic is really working with the unique effects that wood firing can produce. It is what Jonathan Garratt, a potter I am soon to visit, calls 'flavour', a nuanced and carefully orchestrated extraction of very particular qualities. A farmer may grow vegetables to supply a general demand, or she may look to biodynamic, organic or other sensitive techniques to enhance the flavour, texture and uniqueness of her crop. She won't just be growing tomatoes, red in colour and round, but a vegetable whose absolute colour or shape is secondary to its ability to be noticed and appreciated by the taste buds of the discerning consumer.

Sitting with Nic, listening to him and observing his physical presence – worn hands, tattoos and mischievous smile – I knew that this was a personal journey, not one with the buyer specifically in mind, but more of a quest for something mysterious, just beyond his grasp, always reaching.

Nic has always wood-fired, initially because it was his only means of firing. With time, accidents in the kiln, events that would have encouraged him to dismiss those pots, his work has evolved to include these 'mistakes', or 'accidents' as if they were players in the creative act. Decisions abound along the path to create a piece, the choice of clay, its provenance, the shape of the pot, the colour of a slip or glaze, the choice to throw, coil or slab-construct. Usually the firing process is carefully controlled, temperature regulated automatically or by careful attention to the temperature cones, gas or electric heat distributing temperature evenly.

In a wood-fired kiln there are many more possible variants: the more disparate and less regulated temperature control, the constant all-night vigilance, and delegated helpers who need to be in tune. The loading of the wood from the front or side, where the pots are in relation to it, and in more normal situations the careful guarding of those pots from being knocked by the wood or covered in the ash. The wood itself depends on a reliable and economical supply, on being laid long enough to dry so that it can burn efficiently, and on a careful cutting and selection so that it is specific to the size of the kiln for which it's being used. After his years of service to the kiln, Nic has internalised in body memory its temperature regulation. Through vigilance, inspection by means of the apertures behind loose bricks, the colour of the smoke leaving the stack, the heat radiating through the walls, he knows what is happening. He is the cones, the digital thermometer and the clock in flesh-and-blood form.

As Nic walked me around the pottery yard there was a collection of twenty-odd vessels that caught my eye; lying on their sides, upside down so as not to act as rain catchers, they grew amongst the weeds, earth-splashed by rainfall on their mottled surfaces. They looked as though they might be returning to the earth, and I asked if they were a graveyard of hopeless cases. I hope that I hadn't offended Nic; it didn't seem so, but he replied that quite to the contrary, these were his 'special' pieces, the ones that he treasured. Bending down quietly, feeling a little awkward for my insensitive comment, I touched a few and wondered about the language in which they spoke to him, what they told him. As he took hold of a tall, narrow pot some 2 or 3 feet high, I asked him this question: What was it that this pot told him of the time it had been shut in the red-hot kiln?

'A lot,' he replied, running a finger down a paler section of it along the whole length. He said that this was where it had lain in the ash, where through the firing process the ash had built up along it and created this striking striation. His finger continued with the narrative: the occlusions of minerals, prints of scallop shells, varying temperature signatures and the odd deformation of shape. These were not accidents, but the result of allowing things to happen in the kiln that a potter would normally guard against, to harness the wild nature of the process as a contributor to the potter's vision, not fight it or hold it at bay. Yet he did see the creative process as a 'battle', as a war with the forces he worked with, with the balance of surrender to them and mastery over them.

He had spoken earlier of the learning of the potter's craft, of the years it took to master technique and then the lifetime spent loosening that technique, holding the mastery less tightly, allowing other forces to balance the learnt skill. It was this dialogue, this battle, that I sensed kept Nic alive to his work, fluid with discovery and ever evolving. When I got back home, I looked again at the fifteen or so vessels I have of his, and for the first time clearly saw the evolution of process over the twelve years or so that covered my acquisition of them.

The first ones I had bought were clear and symmetrical in form, classical if you like, the effect of the wood firing carefully controlled. As the years progressed, leaves stamped their prints into surfaces, flame scars and ash textures voiced their dissent. The most recent pot I bought, the one I was handling earlier, is born of a riot of interference, appears as if forged from struggle, its textures, occlusions and polished patches refusing to allow it to be tamed by the potter's intent. I see in it the 'warfare' that Nic spoke of, the yearning I sensed in him for a victory that would never be his, a battle that was always lost, but appreciated in hindsight for the small victories within it. When I had found it in the 'graveyard' of my invention, amongst his most treasured items, and handled it, he had looked over and said, 'Good one that,' a look of approval on his face for both my choice and his success.

I am familiar with the feelings he expressed, the sense of excitement yet impending disappointment, the attrition waged by one's craft and the scars it leaves, the fluid delight in a sea of tense hope. Many craftspeople over the course of time have repeated traditional forms within a tightly held discipline of traditional design as passed down through generations. Pots whose forms originate with the clay types and usage, local practices or symbols which imbed themselves in the grain of the objects made. The craftworker in this case is the skilled artisan, the master of the individual movements of their craft, obeying a narrative in which those movements are the words that bring to life a story with roots in a time gone by. There is a rhythm of repetition here, a lack of the 'I', a sense of the maker as part of a continuum of makers handling an old rule book. These are the craftsmen of whom Sōetsu Yanagi speaks in his wonderful essays edited by Bernard Leach in *The Unknown Craftsman*. Yanagi is at times disparaging about the artist-craftsman whose journey is beyond the confines of traditions, into a personal language closer to artistic expression.

When Nic spoke of the seven years, of the imbedding of a skill set, the practice passed down, he saw it as a beginning, as the bedrock from which to explore something beyond. When I asked him of art and craft, of the differences of the two to him, and where he perceived himself to be, he acknowledged that he is on the road from the one to the other, somewhere between. Art, in this context, is the individual expression of the craftworker who has been steeped in process; it is a manifestation of a will to explore the craft beyond the confines of tradition. Nic's forms struggle to free themselves from a set vocabulary, yet bind themselves necessarily to the tension of past association. The clay needs to hold its form, to bind to itself through tension, and the potter seeks to play with the freedom he can allow it while it remains intact to act as a vessel for his exploration.

I find myself reflecting on my own process, on Nic's description of warfare and on a desire for individual expression as a maker. My many conversations with makers have often articulated the subject of repetition, and the monotony and meditation of it. I have often had to repeat actions for the sake of making a single piece. Chip upon chip falling off the edge of a carving chisels, saw cut after saw cut intersecting to allow the release of material, multiple components lying in piles awaiting assembly into a whole that defies the apparent replication that went into them. These actions bore me only if I seek to tame them through time. If I submit myself to their rhythm, I become lost in the action of making them. However, if I am to repeat a piece I have made before, to copy it as if it were from a pattern book and I a bench joiner in Chippendale's workshop, then I become quickly bored and feel enslaved by the tasks, rather than fluid with the possibilities that lie ahead. I am no longer in a flowing creative act, but a repetition of previously practised moves. Is this the point where a creative and developing process of making becomes manufacture?

A conversation I had days after my meeting with Nic further adds to my pondering. Every Tuesday fortnightly I drive the eleven miles from home to the workshop of a good friend of mine, the leather craftsman John Hagger. He initiated this gathering of men a few months before. He had liked the idea of a 'men's group', that was focused on the act of making, rather than on self-development. So he invited six friends of his to come over every Tuesday evening to make alongside him. We spend two and a half hours together and make and chat and make. I have been making a leather bag, while others have made bags, or spectacle cases, stationary trays or even a pair of bellows.

I find that time speeds by, leaving me in its wake, steady in my presence to learning a new craft. I may get to the end of the session, having accomplished little, a hide softened and edges skived, a pattern made or perhaps just a few shapes cut. As an apprenticing leather worker, I am not self-critical, not forever trying to simplify processes for speed or watching the money clock tick away the pennies.

I mention this difference to John, and his face lights into a frustrated smile of recognition. 'That's just the same for me,' he says, a frown creasing his normally smooth brow. He wonders how to distance himself from this and recognises that as maker, manufacturer and businessman he is unable to be in the moment, observing time merely as the hours or days that a job takes. Rather, he marks time by the money that has been allowed for it. How long does a watch strap take to make, or a belt? Is it the time allowed by the fact that there is an order for one hundred at a pre-fixed price, or an individually commissioned one that needs new patterns? The cost is set by his overheads and the time he has estimated, but once set, this time ticks down as if it were the guillotine, as if in coming too close to it, he would lose his head.

There is creativity in each action of making, in its intentionality and in its cumulative growth into an object of utility and beauty. John has become a manufacturing enterprise, employing a small staff and for some time enjoying the challenges of running a business, yet there is a yearning in him for freedom from those constraints, freedom to make exploratively again as if he were at the beginning.

And so we return to Nic, to the seven years and the accumulation of skills and where the artisan then goes in his or her endeavours. Nic learnt from his falling pots, from unpredictability and surrender. He followed the call to trust that his skill set could afford for him to loosen the shackles that bound him to one path, and instead found himself in a wild woods without a path.

John is back at that place, a possible point of departure where the skills that allow him to create such objects of beauty also bind him, the individual, to a limiting path. Before my workshop burnt down, I had made the same choice, having gradually divested myself of assistants, and then of commissions, I was to make for exhibitions only – so that I had time again to explore my craft. Notably, my very last commission was for John, a great bed for him and his wife Debbie, a celebration of them and of dignity. They knew it was to be my last commission, and John puzzled over my decision. Why I would want that and how I would manage? Fair point, I said, but I wanted something more.

After I had delivered the bed, I made some speculative pieces, won a competition, sold the pieces and felt great. Ten days after they were delivered, my workshop burnt down and I've barely made anything since. Life has a way of fracturing the control we wish for and allows chaos back through the door.

I remember the image of Nic by the kiln, knowing it is time to empty it, but allowing others to do it for him, not wanting to get too close to the results as they emerge. Once we allow chaos to get a glimpse through the boundaries of order, we must manage the relationship between the two. Nic will approach his new pots out of the corner of his field of vision, not too close, not too direct. No wonder he feels as though he is at war, for he has let the unpredictable through his gate and now must surrender to its effect on his endeavours.

———

Materials extracted from nature are continually transformed by humans, and continually the relationship is fraught with the unpredictable, and the will to control it enough. Venturing out across Dorset on another of my trips to meet makers, I find myself at the mercy of the elements. I am on my way to see Jonathan Garratt, a potter I had met many years before who specialises in garden ware. My wife Dolly accompanies me, as we are on our way to see friends later in the day. The wind, unleashed by the surrounding chalk hills, gathers over the clay plains and howls around the car as I drive onto the golden gravel of his yard. As I get out, it tugs maniacally at the door and I struggle to rescue it from the wind's grasp. Jonathan is busying himself around the woodpile, now a sixty-four-year-old grey-bearded elfin man, his spirit alive and wild as the wind as he comes up to greet me. He ushers us into the calm of the shop, the wind wildly buffeting the great wooden door as he wrestles it shut and bolts it securely.

To the right is his shop, shelves full with a variety of work, colour speckled throughout, form and content shifting through time. To the left his workshop extends down the narrow building, drying clay, potter's wheels and tools scattered on benchtops. The dusty white of dried clay coats everything, bringing a tranquillity to the scene, defying the bright colours that will emerge from the glazes in the second firing. As we chat, the conversation quickly goes from our past meeting to the present. He speaks of terroir, the idea of place, of belonging, of the connection of craft to the land out of which the materials used for it have grown. He uses the word 'flavour', a nuanced choice to suggest the character of the pots created in this way, where the collaboration of the earth, the maker and the wood-fired process contribute to the texture, tactility and character of a thing as if it were more than just that; as if it were alive, had taken some life from the wet clay and torpid flames that had birthed it.

I have heard another friend of mine, Nigel, speak with the same fervour and language, of terroir, of the expression of the land in the product of it. In Nigel's case he is a biodynamic winemaker, his wines emerging through a light touch, their flavour a pure expression of the land on which the grapes grow. This language is common in wine production, so to hear it used within the context of clay and pottery serves somehow to liberate a narrative, to remember how close the production of pottery is to the source of the clay, and even the glazes.

Jonathan talks of the clay he extracts from the land here, and of its contamination by lime through the ground's previous incarnation as pasture. He speaks of the tiny specks of lime as if they were assassins lurking quietly within the fired pots, ready to strike when the levels of humidity become just right after a little time outside. Lime, processed in lime kilns from limestone mined or imported into the area, would have been used to fertilise the fields for better agricultural yields. These materials, once taken from the earth, were now put back there out of context, and had seeped into the clay layer beneath the topsoil, waiting for Jonathan to gather them up and reprocess them in his pieces. He could not pick them all out, but just had to leave the finished pots outside for long enough to see which ones self-destructed and which ones didn't, and then after a while he could leave the survivors out for sale, confident that they would weather time.

There was a moment when, as we were in the pottery shed, standing by the kick wheel and speaking of its properties alongside those of his electric wheel, that he remembered some French jugs he had seen being made years previously. That each glazed jug standing in a row drying next to the man glazing them had an identical pattern of drips running off one edge, as if purposely repeated. It was only that the man applying the glaze was in such an active and rhythmic relationship with his task, his body so attuned that he repeated the same motions leading to an identical set of outcomes. These outcomes included that repetitive drip.

I think Jonathan told me this story to illustrate the depth of relationship present between the maker and the action he was soliciting with the pots he creates. Repeated often enough, actions become habituated within the body, repeated as if they were the same as the action that preceded them, programmed into the muscle memory and sinew of the maker. The body is, when in regular relationship, able to act as if it were programmed, because in a sense it is. Each action that precedes another serves to create a structured memory within the body that can then be repeated reliably and with ease. In Jonathan's case the small inconsistencies of this process, the failure to be identical, the vagaries of the naturally dug clay and the unpredictable action of the wood-fired kiln all give rise to the unique quality of each pot that emerges. Each similar yet utterly individual and quite distinct to ones produced in a more mechanised fashion.

My wife had been looking at a jug, striped green hues on a neutral background, small marks etched into the surface. It was one of a small collection, all at first glance similar yet, on closer inspection, unique. Handling it, she soon had Jonathan at her shoulder observing the firing patterns on it, the shadows of heat movement, the distribution of colour unevenly through the stacking pattern of the kiln.

Each pot placed will take a mark of its placement, a feature which may get some to speak of 'imperfection' as it is not a marking controlled directly by the maker. Yet as Jonathan's response indicated, it is this very aspect of the unknown that is so attractive to him, that speaks of a certain soul beyond the maker's skill.

Perhaps this is what the Japanese would call *wabi-sabi*. This is akin to the variability Sōetsu Yanagi talks about in his essays, about a certain wild provenance within the practised repetition of the maker of ancient traditions. As a maker works through the principle of risk, engages in his or her practice through the free and practised movement of hands, tools and materials, if there is not a reliance on the certainty of jigs and machine coordination, then chance always has possibility of sneaking through the door. This abandonment to the spirit of a craft is a surrender to uncertainty, to the unknown.

When Jonathan sits at his wheel, a lump of clay before him, or a basketmaker gathers her withies to start a new basket, it is only by means of the interface between their hands and the material that the object will take a three-dimensional form. Whether it is through the actions of twisting, pulling and snipping or those of the hands squeezing, drawing and caressing, the basic relationship is the same. It is a moment-by-moment unfolding of a sequence of practised actions, which respond directly and instantly to the vagaries of the material.

When Jonathan points to the variations on the jug in his studio, it may at first be a little difficult to differentiate it from the other green-striped jugs he has standing next to where this one was. Yet the incised decorations made with a wheel are a little deeper or shallower, the green glaze painted fractionally thicker or thinner, and the shape of the jugs ever so slightly different.

As I talked to Jonathan, my eyes had been scanning the shelves occasionally in the hope of glimpsing an object that might become the birthday present for a friend – the green stripy jug, as it turned out. This jug was not merely a jug, but a very elaborate catalogue of action, relationship and thought that had collected over the whole of Jonathan's working life to materialise in this particular object. Moreover, the clay – processed millennia before from rock that itself had been formed many millennia earlier – was part of a making process so long that we cannot possibly grasp it. That green stripy jug that now sits on a shelf in a small cabin in Devon is not merely an object, but rather a representation of the depth of relationship between nature, time and human interaction that made it possible.

————

This reminds me of the time when I had first learnt that the winning of clay is not something that happens in a competition between keen potters, but rather the digging of clay from the ground. In an age when we take for granted what is beneath our feet, and where we are so used to buying anonymous objects from shops or over the internet, it seems apt that we use the term 'winning' for this application. The occasion when I learnt what it meant was during a walk along a woodland path with two fellow teachers and an entourage of twenty students, ten years or so ago.

We'd been walking through North Wood on the Dartington Estate, the same woods I had walked with Mike earlier in the book. We were on a path amongst a mixed growth of deciduous and coniferous trees having just recently passed a stand of towering *Sequoiadendrons*. I had started off our walk, designed to connect our students with the natural environment that supplied the materials they would use over the coming days. I had made the obvious observations around trees, lifespan, end of life and material usage. The size and grandeur of the trees had created a sense of awe, and evoked appreciation for what I can often take for granted. As the track deepened its journey into the now wetter woodland areas, our feet grew heavy from the mud clinging to them.

At this point my friend Richenda took over the story, highlighting the invisible over which we laboured. She encouraged us to stop and look. To ask why our path was now so wet, and why was the mud on our shoes so sticky? We stopped and lowered our focus to the ground we'd been walking on while distracted by the trees' grandeur. The path was running with water, alive in the rhythm of the ripples on its surface as it moved from pool to pool. The water was a gentle stream, none of it absorbed into the ground, for the clay lining the pools was too fine to allow any permeation. Yellow in colour, bright when scraped back with a stick, but transitioning to redder tones further along the track. We stooped to prod and pick at it, finding it uncontaminated by impurities and smooth to the touch.

Richenda told stories of the clay, of its journey from a primary source where it would have been white, its accumulation of other minerals and particles as it journeyed through millennia acquiring a patina and colour particular to its location. That clay had once been rock was for me such a revelation. How extraordinary that a material noted so for its durability could be broken down through time into a malleable and permeable substance. It was humbling to think of time, the master of this alchemic transformation, stretching so far away from my perception of it as to facilitate such a change. It made me feel so small, so insignificant – a tiny speck caught within the clay beneath my feet, merely one minuscule part of something inconceivably immense.

Pots stand on our shelves, on those of the shops and supermarkets; fired clay is an everyday part of our lives. We drink from ceramics, stand on it; our houses are built from it. As for its fluid origin, the wet clay so often under our feet, we are oblivious of it much of the time, noting it only when the surveyor writes it in a report, and the mortgage company refuses us a mortgage based on the risk of subsidence. Or, when we are digging in the garden, we fight its recalcitrance, curse it for its tendency to seal the porosity of the ground. We can sometimes forget how nutrient-rich clay is, how its ability to hold water benefits growth. In Brazil rainforests are decimated to give access to the pure clay underneath, while in the woods Richenda and I were walking, it's merely a sticky inconvenience on the soles of our shoes.

Years after that day in North Woods, I am again on a walk with Richenda, not far away. Time has swept by, and we are both in a different relationship with our making, sensing the changes around us, the appetite for making that is so much stronger now than it was when we first trained. Richenda runs a ceramic school and has also stepped partially away from her maker's identity. The school has thrived, yet the delicacy and skill of her own tactile creative expression are alive in the pieces she has made, not presently in the process of making itself. Her influence is felt through the school, the work of the apprentices continuing a house style which found its roots in hers.

The walk takes us through paths we haven't trod before, following no itinerary, and finding ourselves on the edge of private gardens and car parks before returning to the footpaths. When we'd shared a coffee earlier outside a local café, a wet northwesterly refreshing us, Richenda used the phrase 'beginner's mind' to describe how she has experienced leaving her old skilled potter self behind. She equates what she is going through to the new potters who arrive at the school, free of any notions of how something should be, open only to what unfolds. When she manages to get into her studio, Richenda doesn't try to make anything particular; she just enjoys getting her hands back on the clay, allowing her senses to engage with the soft capacity of it. She tries to take her own self-judgement out the way and simply enjoy the experience. Like me she is beginning again, and like me she struggles with the unfamiliarity but also embraces what may emerge from it.

Material

As we go through a gate into a great green-brown field, our conversation has left the contemplative and entered directly into the material, the matter of our making and the matter around us. Richenda's absolute passion for her work manifests in the clay and glazes that she uses, and that the studio school encourages and explores. She speaks of the relative ease of tracking the source of her clay and clay bodies, of its origin in clay pits close by, and of its processing in Staffordshire and transport here.

Yet on the subject of the glazes, her passion collides a little with frustration, as she acknowledges how the components of any one glaze can come from all over the planet, their presence as a single colour on the outside of a pot entirely obscuring the complexity of their provenance, extraction and processing. She alludes to child labour, rare minerals, conflict areas and scarcity. She also cross-references some of those minerals to other uses, the cobalt for instance to computers, and suggests that once a material is incorporated into a product, the stories of origin are lost, and the responsibility is buried. A packeted glaze is akin to any anonymous product, its history invisible to the buyer, the narrative now directed by the retailer or wholesaler of it.

She regains her animation, gently gesticulating at the trees, enthused by them and what they represent to her within this conversation. 'They've got all the minerals I need,' Richenda says, and when I look at her a little incredulously, she adds, 'in their ash.' She has to give me a quick science lesson till I really understand what she means. When a tree or any plant grows, it pulls minerals

144

out of the soil, up through its roots, and these become entrapped in the fibre of the tree, within the wood. Only when the wood is burnt do those minerals become available through the ash. Individual plants growing on varied soils will produce ash with different mineral components. Traditional glazes were often ash glazes or salt glazes. The ash was produced locally, creating an individual vocabulary for the pottery, a regional vernacular, an expression of terroir.

Fire is again the catalyst, whether it is for forging tools, for firing the pots or for releasing minerals otherwise unavailable. The fire is traditionally of wood, and the wood of trees, and the trees of a wild growing of the earth, of the clay. Fire allows the clay of the earth and the minerals of the earth to be reacquainted, to be re-formed in a different guise, to become human expressions of what was under our feet. Trees grow from the earth, are cut and used, their branches burnt for ash, which is used to glaze the pots made from the very earth they grew out of. No wonder Richenda expresses such delight. It is a story that contrasts so sharply with that of the branded product, the anonymous glaze that promises absolute results with certainty. These ash glazes usher in the wild, are not labelled with specific formulae, cannot guarantee an absolute outcome.

I work with a material whose provenance I can easily control, whose material character is readily apparent, whose transformation remains almost entirely in my hands. So from my perspective, there is magic in the transformative processes involved in creating ceramics. That a glaze made from ash can be painted onto biscuit-fired pottery to be fired again and in that process fuse to make the porous vessel impermeable and resilient is something of a miracle. It is of science, of reduction and reassembly, of a depth of understanding and rich relationship. That the glaze in its raw state has a colour and texture totally unrepresentative of what it will become stretches the imagination of the maker, and brings me an understanding of why Nic cannot bring himself to open his own kiln. That within the unpredictable nature of his own process is also the unpredictable response of the glazes and materials, whose final form is created only in their direct and fluid relationship with the fire.

We find our way back to where we began, both physically and metaphorically. As we pass by her teaching studio, she veers off and I follow her in. I stand quietly to the side observing students new to pottery, exchanging a word or two with an acquaintance, while Richenda chats to a few people. I see her with Lucy, the same Lucy who had opened the door for me at Nic's studio at the beginning of this chapter. She's working at Richenda's place now, having finished her apprenticeship with Nic. Soon she is to go back home to Australia, new beginnings beckoning on her road to inculcate her passion for making into her growing identity. She speaks of her growing personal style, of the embossed decoration she uses being akin to creating her own character and texture. I hope I meet her again, when her seven years are completed, and when a few more years have folded back in on her.

Before Richenda and I part company, we talk about the unpredictable, about not being in control, of letting go from the seven years of body knowledge and opening ourselves to the trust that something beyond it will emerge. I am left with the picture of the beginner who becomes a professional only perhaps to become a beginner all over again.

Chapter Seven

Tool Hand

A s an artist-craftsman I stand at the tail end of the development of the crafts through thousands of years of the artisan, of the unknown, egoless craftsperson. The heritage that I am in the shadow of has evolved through the Stone Age, Bronze Age and Iron Age. The tools of the early artisans have disappeared, the evidence and heritage of the early period are all but invisible. The line I follow is of only the very recent past, of my tutors and theirs, of a language of making centred on a post-industrial aesthetic. I have seen myself as a handcraftsman, as a maker who uses relatively simple tools and techniques, many unchanged for generations. Yet in all truth, my lineage is that of iron and steel, of industrial thinking and relatively sophisticated technological development. For much of my working life, I have taken the tools I used for granted, as I did my own ability to handle them.

The writing of this book has passed through the two-year anniversary of my workshop fire. The workshop was fully rebuilt nearly a year ago; the interior shows some scars of the burning, the outside is all fresh wood and shininess. I built new wooden workbenches recently, ordered all the cleaned teaching tools to pristine cupboards and am in the process of putting the finishing touches to everything before the courses start again. It looks great, sparkly, shiny, no sign of black. Yet there has been a niggle at the back of my mind, an itch I wanted to scratch but didn't dare unless I scratched it raw. I have a cupboard in the machine shop, a tall, narrow space packed neatly with stacks of dark grey boxes, each labelled. Amongst those that say MACHINE PARTS, FIBRE PADS, SPARE SAW BLADES are a collection of large and small ones labelled as PLANES, OLD WOODEN TOOLS, BLADES and so on. I have seen them on my periphery for the last two years as I have rebuilt order around them, yet I have left them alongside the open buckets of other filthy components where they are.

Recently my hands pulled me to the cupboard, dared me to bring out all those boxes and take off their lids. Inside were the half-rotted remains of tool corpses, and I started on the long and laborious process of bringing them back to life. Two years ago friends had come to help me after the fire, a retinue whose hands nurtured the most precious tools, cleaning them as they sat round tables chatting in the soft autumnal light. Those tools that were not a priority went into the buckets and boxes, a conscious desire for them to be 'out of sight and out of mind'. The cleaned tools have re-rusted, have goaded me in my effort to move on, have moved me to comment continually that my carefully kept tools had never had rust before. Life is not neat.

Tool Hand

Tapping at the keyboard, my fingers extensions right now of my thoughts, I see them blackened at the ends, traced through with dark lines etched into the skin. They speak of iron tools cleaned and sharpened, of the oxide leaving them to lodge in my skin. They speak of the fire, of two years past, of the job I have dreaded and delayed till the right time. So, these tapping hands hold trauma, yet I move beyond it as I have started to heal those neglected tools. The soot that burnt onto them from the atmosphere of heat and flame lies once again in my skin. The smell that occupied the breathed air of my every moment now casts a faded memory into the breeze that has blown the rest of it away; its sweet acrid shadow rests on the edge of my consciousness, and I wonder at the accuracy of my scarred memory, which remembers it as bitter tar, resinous and cloying, stuck into the fibres of my clothes and the filthy strands of my neglected hair.

There are six or seven large plastic containers on the floor, four of them almost too heavy to lift. They are filled with hundreds of soot-patinated tools. Old wooden planes, spokeshaves rusted red-brown alongside multitudes of matching chisels and other blades. There are compasses and scrapers, saw sets and saws that are rusted beyond salvage. Wooden handles blackened at the centre with soot, char towards the edges, a heat signature reminding me of where they once were stored. Looking up into the white, dry and clean roof of the workshop, the bright lights and orange floor, I remember the darkness, the horror.

The rust, soot and tannin etched in my skin are vestiges of a scream. The sweetened smell of it blessed by time is no longer full of horror, only of the promise of resurrection, of what rises, of the phoenix bright and clean angling its flight towards the sun. The wire wheel, wire brush, waxed caress burnishes the past back into the past, just a shadow of memory scratching at the synapses. I wonder at the glow that surfaces, the burnish of the wax I have buffed into it, feel a lack of trust in it as if it were not real. It is the now, the *maintenant,* the present moment of attention, the maintenance of care which helps resurrect memory afresh. I have not enjoyed the actual job; hundreds of tools and blades picked, dismantled, scrubbed clean, waxed and reassembled. I don't like the black filth imbedded in my skin; I don't like the darkened water running down the sink's edges. I don't like the smell. It is only when I write that I understand the beauty of it, that the actions that allow me to clean the tools have an effect so far beyond the physical. Somehow the manipulation of the tool, its articulation in my hand, helps heal me, sets my teeth, fettles the cutting edges, and brings me fully back into my human purpose and potential.

An emptied bucket fills slowly with bright flashes of silver and bronze, cleaned tools bedded down neatly awaiting their future. It is not about the need I have of these tools that has driven me to clean them. Those that were important were cleaned long ago. I have cleaned them simply because I couldn't leave them as they were. They were neither living nor dead, zombie half-life with no purpose in either. That they shine is good enough for me, that they are now fit for purpose, whatever that will be. To be gifted, sold, used or tucked away. They are tools again. No longer rusted iron and the green oxide of copper flushed bronze.

I wonder what they mean to me, these tools, tools at all. I would not be a craftsman without them. Wood would be a stranger, not a friend. I would seek connection only as an observer, unable to partake, to disassemble so that I could reassemble. My fingers would itch from the fallow ligaments and muscles of flaccid hands. They would itch for purpose, for meaning, for the metal edge that would make them significant, make me significant.

A tool is an extension of a hand. My tools are an extension of my hands. A rusted tool cleaned and repurposed becomes an extension of a hand new to its use. Our hands have grown with tool use, are now as if always designed for it. The tapping of my fingers here – a shorthand for their potential. I look at my hands and know they are not hands alone, know that their potential rests within the articulation of the body they are attached to.

As my fingers hold the blade, I sharpen, twist it in the light, my forearm facilitates the pivoting of my hand and of the tool to catch the light. As I lift it further my shoulder flexes, the muscles contracting, the potential of my hand rippling through my upper body. I manipulate the tool, my *manus* holds it and the action I practise facilitates the life force of the tool. I, a human, manipulate my destiny through my interaction with my surroundings, and tools allow me to do this potently.

The tools lying in the boxes were made by the hands of others over the past two hundred years. I have touched iron forged in 1826 and in 2017, in Bristol or in Japan, made by skilled smiths or in computer controlled factories. As I have cleaned I have watched tool design change alongside the working methods. I have seen the fold of the melded metal struck by a craftsman's hand to join soft robust steel to the hard cutting edge that wood desires. I have witnessed the transition of a local craft to an industrial practice. I have carefully avoided the pointed and sharp teeth of the saws: those giant with their purpose as tree-felling saws, nearly as tall as I am; those tiny and delicate Japanese saws whose configuration owes its form to an entirely different thinking. The westerner pushes their tools over wood, the Asian traditions pull theirs, the tool finer, its resistance less. I push my way through the cleaning and sharpening with such fervour I don't notice that I have scraped my hands raw and their blood patinates the tools.

Gabriel, who is apprenticing with me, is on the other side of the workshop, opening a cardboard box newly arrived. He coos with delight as his hand emerges from the box clutching the bright light of new steel. While old tools are resurrected in the filth of flying oxide and metal shards, he spreads out a growing selection of new chisels and saws on one of the newly made workbenches. There is something seductive here that pulls me to leave my task, yet the enthusiasm I would normally have for a new tool is tempered. I am pulled more to the restoration of the old than I am to the new. They remain in their boxes, waiting for my scrutiny later.

When I do get to them, their pristine factory shine alerts me to the filthy state of my hands, and I retreat to the sink, shamed to abrade my hands as clean as I can. Returning to the tools, I unpackage those still packaged and spread them out. They are new tools for the teaching sets, supplements to what I already have, high-quality tools from Germany and Japan produced mainly by machine. Notwithstanding their shine, their regularity of form sets them apart from the old tools I have cleaned, those showing obvious signs of the handmade. I doubt that there is a superiority of one over the other, only the narratives they carry setting them apart. While aficionados of old tools talk of the high quality of old steel, those of modern tools speak of the reliable production techniques, absolute hardness on the Rockwell scale and quote letters and numbers of the steel type, which is designed to imbue the prospective buyer with an awestruck confidence.

There is a growing culture of tool obsession within the woodworking community, driven at least in part by more people drawn to try spoon carving, bow making, furniture making and so on. The enthusiasm has spread into the toolmaking culture, when old crafts are being relearnt and old tools either resurrected or reinterpreted. While in the postwar era, tool production was blighted by a shrinking market for hand tools, many designs being discontinued and others made ever more cheaply; now there are websites specialising in handmade, expensive tools, narratives abounding on the perfect tool form or the best maker. You cannot be a self-respecting maker now if you don't own a particular plane or chisel or spoon carving knife. Tools have become identity, a mark of self-respect and a signal to others of the respect their user deserves, as do their associated skills. At the same time the branding industry moves to identify its corporate clients with the 'handmade', the 'crafted', 'artisan' and otherwise unique association to old traditions. This is irrespective of whether the products are made in workshops or factories, or whether there is any truth in the assertions.

Amongst all the rusted tools are many old planes, wood-bodied tools, their blades rusted hopelessly into the iron blue stain on their Beech and Oak bodies. Some are up to 150 years old; most at least predate the First World War. Some belonged to my grandfather. Their grandchildren are already sitting cleaned in pristine cupboards, their iron bodies and shiny caps prioritising them. As grandparents fade, so, too, does the memory of them as anything other than grandparents, that they were once young, had dreams, felt up to date and contemporary. The old grapple with their place in a changing world, and so, too, will these tools. They are anonymous tools of an unknown maker. Only the blades ever hold a maker's mark, that of a local smith, the bodies only ever having the carved name of the owner to identify it as his. I am undecided for their future; some I do occasionally use, but if I hang onto them it will only be for the stories that they can help me tell. If I sell them or give them away, they will sit on shelves somewhere, perhaps storied back into life occasionally, otherwise just gathering dust.

They have come out of the boxes, and lie disassembled on the benches. A wood body, steel blade and cap, wooden wedge and perhaps a brass nut or foot if they are more complex. The steel use is kept where it is essential; in the blade. Before centralised production and distribution, with wood easy to source locally and metal expensive, this was the norm. Steel-bodied planes had been around since the Romans, the technology for casting their bodies quite accessible, but the facility for their production and maintenance much less so. My planes tell the story of society's changes, as their design shifts as they pass through my hands. I have cleaned all the wooden bodies, and the heavy metal blades that went into them. The early blades are tapered, hard steel forged to soft, a fair weight in my hands. The bodies are carved out of Beech wood, sometimes Oak, and for one set of pattern-maker's planes I have, Boxwood, slow-grown, some forty years across its tiny width. With time, the steel becomes thinner, straight and not tapered, production streamlined for convenience rather than faithful to tradition.

Through the 19th century, more metal-bodied planes emerged, first infill planes with metal soles and wooden bodies, and then gradually the steel use increased until only the small handle and front knob remained of wood. I have spent a lot of time during my life in the foundries of the Midlands and Sheffield where the Industrial Revolution found much of its expression. These were the places where those metal-bodied planes would have been cast alongside lamp posts, bridge components, machine parts and just about every other conceivable piece of the puzzle that defined the industrial era. As a young designer and manufacturer, I journeyed to and from my home in London to the workshops of Sheffield, Birmingham, West Bromwich and Walsall over a number of years back in the late 1980s and early '90s. I watched molten steel, aluminium and bronze pour into the very same moulds that it had been pouring into for hundreds of years. The foundry I used most was a great cathedral of a space, cast iron columns and superstructure born out of the very techniques being practised within it. Scattered across its great floors were groups of workers, trades and skills separating them.

The pattern-makers had a space separate to the bustle of the main foundry, the tools and material they used being those I was familiar with. They made the shapes of the items to be cast from wood, but now their skills have been taken by machines and computers, their specific tools relegated to mantelpieces. Drawn to the idea that I could use my woodworking skills to create my own designs in metal that would not have worked in wood, I had developed some of the specialised skill set of their trade. I was taught in trade schools that still existed in the '80s by men who had come through the tradition of historic apprenticeships. When I walked into the great sand-encrusted floor of the foundry, I always felt like I was a pretender. They called me 'lad', though I was thirty and they were only a little older. To them, of course, I was a lad, a youngster in terms of my experience, a pretender in their midst. I don't remember understanding what it was I was witnessing in that cathedral other than the wonder of the transformation of molten metal into the image of my imagining. I missed the cue for the end of an era, barely realised that what I was privy to was a huge privilege, part of a heritage particular to that area, and soon to be swept up with economic change.

Tool Hand

The sand on the workshop floor was for the casting process. Steel forms were filled with it, one or a pair for each piece to be cast, and into them were pressed the patterns for the objects to be made. A plane body, barely recognisable in its pattern form, would have been pressed carefully into the sand, a channel furrowed for the passage of the molten metal, and the top box placed on with a pouring hole pierced through. Even the steel plane needed the skill of the woodworker, yet now encased in sand it was in the hands of these men of fire, with their sweat-red brows and great leather gloves. The crucible of flaming magma was poured skilfully into the opening, at a speed yet with care. Rapidly they would have gone from form to form, prepared ahead of them by the other team. Knocked out when cool, these crude shapes lay strewn on the foundry floors, soon to be gathered and taken to the fettling room, where they were ground and cleaned up, ready for machining.

As the hands of our ancestors went from gripping branches to tools, the path to our present form as humans was set. I grip my tools, whether of metal or wood, and with my grip and the dance of my body around it, I advance my destiny. I am wild man holding a napped rock, using it to shave off the edge of a branch to make a spear. I work out how to lash it to a stick to make an axe, and maximise the force that I can apply to it. I am of the Bronze Age, extracting copper and tin, creating enough heat to melt them, and with them make the axe head anew, keep it as sharp as it can be, and find in it new skills. In the Iron Age, I find my modern counterpart, the axe head, becomes chisel, adze, hammer, plane iron and saw, and with it I fashion the wooden body to hold it in. The saw teeth emerge and slowly the struck blow of the axe used to cut appears unsophisticated and crude. Saws become ever more the norm, the sophistication of their manufacture developing through time. The stone becomes the sharpening tool to keep the iron edged, and I find myself with a workshop full of sophisticated tools, while at heart I am the man with the rock shaving off the bark from a stick.

Within the historical and prehistoric periods we are so familiar with, the Stone Age, the Bronze Age and the Iron Age, there is a simple tenet. We have been defined by the materials from which our tools and devices were made. Or at least by the materials that carried the edge of our intentions and potential power. The Stone Age gave form to the napped stone, napped on itself to give it an intentioned edge with which it could become weapon, tool, axe head or hammer. The mastery of high temperature and technology of metal extraction from rock seams gave us first the Bronze and then the Iron ages, defining tools again, spearheading trade and establishing the wealth and prowess of groups.

Yet as my boxes of tools describe, these materials that defined ages were held by a material that was given no age of its own, that was always in deference to the power that they held. Wood is the unclaimed hero of all times. It needs no digging for, grows back when taken, and has worked with the three early material ages. Whether as handle to a stone axe head, fuel for melted bronze alloy, props for the pits and tunnels, or bodies of planes, it has been the facilitator of human development. It is the material that has claimed no absolute identity, that has been stripped thoughtlessly from the landscape and that is now under threat from new diseases and the effects of global warming. It rots back to the earth, leaving little sign of its extensive use in tools, while the rocks, bronze and iron components are dotted in museum collections the world over, claiming their place in our evolutionary history.

My planes sitting in those boxes are not ancient, were in common use until quite recently, and their Japanese, German and Asian cousins still are. Their genetic history goes back at least to the Bronze Age, where the fashioning of crude blades required their housing on wooden handles. So much has changed and yet so little.

The world is full of limitations and it is our ability to know whether to respect or push through them that helps define the life we live. Submission plays as important a role as dissent, and as a maker one can somehow be in both at the same time. Clarity of action is a vital component of the maker's craft, yet so is submission to the natural forces of the materials and tools one is working with. To be at once decisive and submissive is the art of much making. Machines can alter the relationship with resistance, can still submission and create the illusion of dominion. The dominion is not of the individual maker, but of the machine itself. It ends up dictating and stills the pattern of relationship otherwise open to the maker.

All tools are to some degree machines. They are all adaptations of material to enhance the productivity of the hand, of the human. A knife extends the faculty of the hand to cut, kill or puncture in a way that the hand alone can't. The stone that is napped, the bronze that is cast and the iron that is forged are all material extensions of the faculty of the hand's relationship with our environment. The knife in its earliest forms allowed a huge developmental leap in the ability of the human to interact with the natural world, and to develop dominion over it. Each tool development since has been a refinement of that dominion. We create a version of freedom through dominion – ours over the natural world. Yet what do we lose in the process? Is there a tipping point over which the benefits to us of a tool become tilted away from us, and begin to nibble at our very core?

Leaving my thoughts and the piles of cleaned tools, I head from my hand shop to my machine shop. My machines are all relatively old, some my age, some a little younger. They, too, are shining in their new green paint, belying their age and the state they were in not so long ago. With big electric motors, a great weight of cast iron and a screaming vocal urgency, they make me more powerful than I am. They allow me dominion over the wood I process, allow me a speed I could not otherwise have, and through their use I run the risk of becoming their prisoner, a subject of their urgency. With them I can cut, flatten, mould and curve great planks of wood ready for the more delicate and subtle role of my hand tools later. I respect these beasts greatly, have scars to prove my allegiance to them, yet these days, before I habitually go to them, I first check in to make sure that I am not better doing that job by hand. I have grown ever more aware of the dust, the noise and the vibration, which are the consequence of their efficiency, and long for the silent breath of a sharp blade as I push it by hand over a piece of wood.

———

With this and the blades of my planes in mind, I find myself opening a gate onto a small field bounded by a neatly cut yew hedge. The field holds a promise of the wild within it, of something more spirited than the conformity of its boundary. The clasp is hand-forged, and so tightly latched to the post that I can barely undo it. The path runs alongside the field down towards a small wooded area and another gate. The damp woodland call beckons as I open the second gate, a carefully carved sign on it saying *PRIVATE* and the insignia of a hare contextualising it. The mark of a maker welcomes me to a private world, one of invitation deep within the wild growth, far away from the carefully tended yew hedge.

I am within a maker's space again, far from my car parked half a mile down the road, far from industrial estates and main power cables. It seems that the makers I am drawn to are often those who seek the margins of the wild and the tamed, those that, like the scurrying and flying creatures, seek the shelter of the hedgerows. Once again I am in a woods, having chosen the maker within it, and he is the one who I seek here somewhere within the shade. Stands of trees merge as I walk down a well-tended trail; separated into groupings, they morph from Ash to Birch to Hazel and then to Oak, the leaves on some of these quite different from the others, larger and more pointed, their trunks unusually straight. Common Oaks hybridise with Turkey Oaks, their children all unique combinations of the character of each.

Where the track divides into two paths, there stands a tentlike shed, scars of tractor tyres on the soft ground within it, fuel cans close by and 50-gallon barrels of diesel sitting outside.

I skirt close to the boundary of the woods, a line of Ash and Alders keeping me in the shade, great furrowed ditches cutting through it to drain water off the land. Ahead of me and to my left, the dark shadow of a great rusted iron cylinder rises 6 feet out of the ground, and even wider in its girth. Round it lies scattered remnants of charcoal and ash, and on one side a pile of cut branches accumulating for the next firing. It is a charcoal burner, a more modern way of depriving slow-burning branches of oxygen to render them down into slow-burning charcoal. I imagine that this is the fuel for the forges, carefully prepared from the woodland harvest, the autonomy of the maker from his own woods.

I branch intuitively off to the right, looking through the trees to see if I can see a building. I glimpse a roof and as I follow the path towards it come across a large quantity of propane cylinders grouped together as if just unloaded from a lorry. Wild wood and bright orange cylinders sharing community alongside the red tractor diesel and birdsong. The charcoal burner's significance becomes hazier, a more complex relationship clearly afoot than the one I had imagined. The building is now fully in view, with the sound of a generator pulsing from the other side of it. There is a small red tractor parked alongside the workshop, set in neutral with its PTO (power take-off) driving a generator. I open the door ahead of me and find a small workshop space, two lathes and various other equipment around it. The bright colours of machine brands abundant but nobody there.

I close the door and move around the side of the building to another door. As I approach I hear the pressured roar of propane and on entering feel the warm pulse of air from it. Dave Budd stands in front of a small furnace, a pair of tongs in one hand holding a glowing piece of steel. He sees me, acknowledges my presence and asks for a minute to finish what he is doing. He spends a little time twisting the metal into a spiral, sparks dancing off the edges of it, and then puts it aside and turns the furnace off. I have watched him as he stood intently emerged in the task, aware that my presence here was an intrusion in his otherwise solitary working environment.

A lot of the craftsmen and -women I have visited work alone, and often in unique environments. I have unwittingly collected a grouping of humans connected through various common characteristics. Yet in the woods with Dave, I felt that I would feel isolated, not just alone. His company was his tools, his kit, the thrum of the tractor and the embodiment of the energy present in his work. The workshop was full of possibility, of tools and machines that could together help him fashion any idea that his mind sought to grasp. All this machinery and industry within the confines of a wild wood, invisible to the world just a stone's throw from it.

As we talked, our mutual interest in the historical context of making soon moved us to another shed adjoining the first. It housed his book collection, leather-work facilities and other cleaner needs. With an archaeological background, Dave is fascinated by how tools would have been made, and the tools that would have existed for particular requirements. Those he makes are often for specific purposes, for the makers of buttons or of clogs or of trades and purposes all but forgotten. He takes a great book off his shelf. Written by Denis Diderot in the 18th century, it is part of the great work *Encyclopédie* he produced over a twenty-year period. It is full of engravings, detailed images of craftsmen at work, of their tools and the objects that they make. Tools and machines blur into the same object, their purpose to assist the maker. The book was radical, as it was to popularise the crafts and the process of industrialisation that was transforming them. It allowed people to see what was happening behind closed doors, and opened the era of machine development to the public. One engraving exposes the workings of a small workshop, three or so makers at work on various tools. It is a button maker's shop, full of wood offcuts being reduced to tiny buttons, the methods spanning the ages.

As one maker sits on a bench using a bow drill to make the tiny holes, another uses a large wheeled bench drill to drill larger holes. That device is of the picture's time, while the bow drill in its primitive simplicity harks back to the Stone Age. They are both appropriate in that context for the tasks they enable. Today the picture is different, these processes relegated to the workshops of a tiny few, while the mass of production happens in the whirling anonymity of algorithms and flying cutter heads.

Dave is approached by makers who can't find the specific tool for a task, who still relish the challenge of making by hand, resurrecting old skills and copying old products. He relishes the challenge of researching the tools that would have been used, and he works to recreate versions of them appropriate to the craftworkers commissioning him. He was full of the enthusiasm of discovery, showed me early efforts of experimentation and a broad selection of the knives, chisels and bespoke tooling he has made. I sense in him his purpose as a maker, to unearth some of what has been lost, to inculcate into his own skill set what he is able to resurrect. He is bringing his archaeological interest alive, creating a living archaeology, a representation of past processes that may influence our future relationship with tools.

Later we are chatting about the woods, his purchase of them years before and his relationship with them, with the trees, the drainage, the spirit of the place. I ask about the charcoal burner and the propane, about sufficiency and reliance. It turns out that the charcoal is mainly for courses and re-enactment, for either recreating the way things were done, or creating an atmosphere and association suitable for teaching. The charcoal heat and nurture necessary make a great adjunct to the experiencing of a skill, the embodying of it. I reflect that the apparent contradiction of Dave's use of propane and diesel for his commercial work, with the charcoal for teaching, is similar to my own practice.

Material

I rely on a wide variety of tools and machines for my work, yet I teach almost exclusively through hand tools. I use green woodwork as a teaching method more because of the learning it allows for than the transfer of skills. The simplicity of the actions helps engage and support learning, the tools' actions and demands being easy for a newcomer to assimilate. Joinery skills, machine tools and the exacting measurement they require act as potential impediments, reflections of difficulty and resistance. The use of charcoal engages learning very three-dimensionally through the transformation of wood into charcoal and the nurturing of its heat to forge. It takes us back to the origins of heat, to the earliest transformation of metal from ore to object.

———

My thoughts take me back to the witnessing of metalwork without these contradictions, to the pure heat of wood fire and cooperation of community. Between 1993 and 1997, I lived with my family in Mexico in a town called San Miguel de Allende, 2,500 metres above sea level and three hours north of Mexico City. I was making furniture from a simple workshop on the outskirts of the town, while my wife taught art and embraced motherhood. It was a beautiful time for us, two of our children born there, and the first soon bilingual, having left the UK as a two-year-old. The Mexico we lived in was textured with creativity, full of the handmade and an indefatigable sense of adaptation and make-do. The plates we ate off, the lampshades over our heads, the chairs we sat on, the napkins and tablecloths, the bathroom tiles and the bricks from which our home was built were all made by hand, most of them very close by. The spirit of indigenous and colonial craft was still alive and strong.

When life gave us the chance to engineer a break for a few days, it was to the state of Michoacán, a fourteen-hour bus ride away that we would head. Of all the Mexican states, this was where the making culture felt most alive in the streets, in the noise of mallets on wood, hammers on copper or the rasp of carving tools. Vendors had stalls full of brightly coloured carved animals, hand-embroidered cloth and raw, unglazed ceramics, the mica shining out from the dull clay base. Venturing out from the town of Pátzcuaro, where we always stayed, to neighbouring communities was a pilgrimage of craft that we loved to follow. The kids were young, yet in the spirit of Mexican culture, we all travelled out together, street food, vibrant street life and the open-heartedness of the locals keeping us all engaged.

I can hear the rhythm of pounding sledgehammers keeping time with my memory and recollection. *La comida* is at an end, *la siesta* is coming to a close, the silence of the community is again punctuated by the sounds of industry. The streets ahead are full of the ringing of steel against softer metal, the shopfronts filled with the shine of copper vessels, and the grackles are in full afternoon chorus around the bandstand in the village square. This is the town of Santa Clara del Cobre, St Claire of Copper, one of several towns in the region focusing on a single craft, its history parallel to this craft, its story rising out

169

of the conflict between the indigenous peoples and the Spanish invaders.

The story goes that the repressive actions of the Spanish administrator Nuño Beltrán de Guzmán had driven the locals to the edge of a revolt, and the Bishop of the State, Vasco de Quiroga, was forced to intervene. The indigenous people here were the Purépecha, and prior to the Spanish occupation they maintained strong traditions of artisanship. Quiroga realised that by supporting their crafts and further training them, he would create more peace, and compliance, in the population. Through their craft skills they could trade and support their income, and in consequence create a degree of autonomy for the populace. So Santa Clara became known for copper work, Paracho for guitars, Tzintzuntzan for pottery and Nurío for woollens. Other towns for other crafts. Some of these products would be exported to Spain, further developing the potential market. I surmise that the effect of this intervention went beyond its intention, that the singular craft which defined the towns brought to their inhabitants a sense of individual purpose and autonomy as well, that the twinning of hand, tool, community and livelihood was rich beyond imagining.

The process of making copper pots that I have seen in Santa Clara is akin to a performance act, a cooperation between fire, humans and copper that was traditionally mined and smelted in the area. On one trip we had taken my mother to the town, and she wanted to buy us a gift. We parked the car on the long main street, and while the kids danced excitedly around the stalls selling traditional chilli, tamarind and peanut sweets, we tried to find a workshop that we'd been to before, a family business like all those here. Entering the shadows of a passage between two shop fronts, and following the ringing of hammers and trail of charcoal smoke, we arrived in an inner courtyard alive with the texture of transformation. Four men stood around a brightly flaming fire as a young lad stokes it and fettles its energy. They rest on long-handled hammers, chatting relaxedly, having just paused from their work as we came in. The scene is ancient, sketched with a sense of industry, but essentially of family and community, of life and cooperation.

There are multiple hues forming a complex palette, the pigs and hens, a tethered goat visible through the haze of smoke, children playing, and mothers and grandparents busying themselves on the edges. The maestro Juan comes up to us and we greet each other, exchange news and introduce him to my mother. I excuse our intrusion and ask if he and his family could make a pot for us as we watch.

We move from the smell of woodsmoke and the pall of the shadows to the sharp shine and cleanliness of the shop. We move from the dirt and endeavour of making to the silent display of polished objects and the depth of skill and time etched into the contours and crevices of hundreds of platters, pots, vases and basins. We have a look at the various pots, decide on the shape and size of a vase, and chat with Juan about what we want. I am conscious of the imposition on him, or rather invent it as an imposition from my own cultural context. I limit the labour and time by going for a simple design, stilling my desire to see them execute some of the complex detailing that defines much of the work, nervous also of burdening my mother with too much cost. The pots and vessels are often very facetted, indented with repetitive protrusions, infernally complex to imagine the making of.

The ringing has continued in our absence, one of Juan's sons standing in for him, and as we approach, they are just finishing off their work, the hammer blows more delicate, the pace quite slow. We find logs to sit on, while the kids stand, fascinated by the visceral texture of the image in which they find themselves. The fire is being stoked again, and one of the other sons picks up a rough copper ingot and throws it on. The bellows are pumped by foot, inflating lungs driving up the heat, silent orange-red sparks drifting and disappearing into the air.

Time stills and moves quietly, and without speaking, the four men move around the anvil, the youngest son crouched by the fire holding and turning the ingot with long-handled tongs. He brings it to the anvil and places it down, and without a word, the great hammers ascend in order like multiple second hands on a clock face. It is a performance that remains with me, so beautiful and engrossing to watch. As one hammer hits the copper ingot and rises, another is already falling, and in the tiny interval, the copper is moved fractionally towards the held image of its future form. The kids are enraptured, quite silent, their eyes following every movement, every clenching and releasing muscle, every indentation that appears in the copper's surface, every rotation of the clock's hands.

The copper form rises from the ingot, the tensions in it supporting a new shape as it goes from fire to anvil and back. The red-hot and curved plate it had now become is placed on a new anvil, its round head mirroring a sense of a future shape, the hammers raining down on it encouraging it into malleable compliance. The men's faces are flushed with sweat, the repetitive actions demanding their energy, the heat of the day still alive even though the sun is now low. The cooperation of the men, the unity of the family, as the women are always close, glasses of *agua de Jamaica* brought to us, a stint on the bellows or tongs. Clothes dry on lines away from the smoke of the fire; small children play at the rim of the scene.

The copper is hollowing, stretching itself around a bubble of air, a new long-necked anvil reaching inside, and three of the men have now gone back to the first anvil, while only Juan rotates and hits the vessel. The neck slowly closes in on the anvil, the body bellowing out from under it, the edge forming as a lip around a mouth. The copper is touched by a signature of heat, rose reds and golds striating its surface. Juan again asks me if I am sure that I don't want him to decorate the basic form, and when I repeat myself, I sense his puzzlement, disappointment even, that I am somehow stymying his creativity. So he moves to burnish it with a different tool and longer caressing actions, as he works around it repetitively. The colour and shine of the heat-hewed copper emerge gloriously, the texture of the hammer blows like scales on a leaping fish.

This vase sits alongside another decorated one on a shelf in our kitchen, a great ceramic *olla* from Pátzcuaro standing nearby. There are Mexican bark paintings on the wall, and various other objects we collected on our travels. The copper work is redolent with the story of its making, of the cooperation that brought it into being. The persistent adaptation of the human hand to an ever-greater vocabulary of skills has paralleled the development of community and language. As human groups gathered, their hands busied themselves with making and hunting, and language developed to help facilitate cooperation. As I leave Juan's *taller* in Santa Clara, I hope that the scene I described is still alive, that the demand for copper pots and the robustness of community cooperation are still as we witnessed them then.

I flick back over the pages of Diderot's *Encyclopédie*, over the countless images of industry in it. 'The mechanical arts', as he called them, are alive on these pages, the exploitation by man of all aspects of the human world, including the human, are exposed. The ingenuity of handwork unfolds in the great machines and furnaces, the sugar plantations and exploited workers, the great mine workings and the row upon row of looms in great cotton-clouded factories. There are artisans making guns and cannons, swords, cutlery, locks and tools. Amongst them is the image of a copper workshop, artisans crafting vessels by hand, beating and riveting and tin plating. It's not so far from the image of Juan's workshop, but there are no children or animals or women; there is a communality of men and hard work, and not much else.

Chapter Eight

Heritage

In her book *Surfacing,* the poet Kathleen Jamie speaks of a trip she did to Alaska, where she witnessed the effects of the melting permafrost and the land losing its grip to the onward rush of the sea. Perhaps we, too, have lost our grip, much as the land through global warming is losing its own.

The grip that defines the newborn's entry into the world, that with which they claim their belonging. The grip a craftswoman brings intuitively and fluidly to the tools with which she crafts tools and objects for herself and others; the grip that these objects have on our culture and our lives. Our hands are changing their integral structure as we spend our time tapping and scrolling instead of grasping, twisting and articulating them. As we lose our grip, perhaps we, likewise, come to lose our understanding of the connections that bind us to this life.

The land of which Kathleen Jamie speaks is that of the Yup'ik peoples at the village of Quinhagak in the circumpolar North. As sea levels rise, the 2 or 3 metres of tundra at the edge of the water gradually fall into the sea, releasing into it all the objects held in the permafrost for hundreds of years. Objects that even the elders have never seen, objects removed from the material lexicon since the missionaries came to this land and shamed the indigenous people out of their rich heritage. Torn between leaving them be on their journey into the invisible depths of the sea and collecting them to preserve the stories they carry, the Yup'ik elders finally allowed for collection and cataloguing. The result, beyond any ethnographic or archaeological value, was to help reunite the peoples, particularly the young, with the stories they had beheld only as a shimmer. The objects – line weights, harpoon heads, jewellery, wooden arrow shafts, fishing weights, darts, models of animals – were made of found materials, crafted amongst the peoples as long as they had had existence as a people. These objects had helped define their culture, brought to it the particularities of their landscape, the indigenous wildlife and their habits.

As the missionaries came, and the traders arrived, Western goods gradually pushed out the indigenous ones, and the ways and beliefs attached to them. As these artefacts re-emerged from the ground, the young could see them for their heritage, begin to understand their environment in relation to them, and find in themselves the resources to replicate them or to be moved by them. They could begin to reclaim a relationship to their ancestry long robbed from them.

The landscape has historically revealed the materials that have helped define the life of those that live there. The tribes of the Pacific Northwest in North America, who relied so exclusively on the great yellow and red Cedar trees, were defined through the capacity of the tree to be transformed. The bark, the boughs, planks split and fibres stitched. They refashioned the tree to form their own built and lived environment. From roofing shingles to clothes to canoes, the tree provided warmth, transport and shelter. For the Yup'ik I can only imagine the feelings that arose as this heritage of theirs was regurgitated by the melting land of the past into the foaming waters of the present.

Do we all not possess or value an object which speaks of something beyond us, something to which we can connect ourselves but struggle to retain direct experience of? Museums are filled the world over with objects from civilisations whose proof of existence remains entombed in them, whose voice is articulated only through translation. All civilisations are built on the bones of entombed earlier civilisations, the cultures destroyed, their artefacts now preserved for posterity. Nations fight other nations to restore to them the material heritage stolen in the past. They need to reclaim their stories to reclaim lost identity. The Parthenon friezes remain in the British Museum, and with them some element of the identity of the culture that created them, and the modern culture ceded from it.

What of the makers, the men and women who left their marks in the amphorae and burnt bedsteads of Herculaneum, the frescoes of Pompeii or the arrowheads on the beach at Quinhagak? What of the life that those objects gave relationship to? Peoples, families and individuals have come and gone for millennia along with species, cultures and habitats. Often all that remains are the stories, preserved either in oral or written form, or through the artefacts left behind or ones that re-emerge. A new skeleton discovered on a mountainside entirely changes the understanding that scientists have of our evolution. A rock with a particular tool mark speeds up developmental time by thousands of years, and a shaped stone under a Christian church speaks of a culture and way of life totally obscured.

When my grandmother returned to Vienna in 1946 after surviving the concentration camps, she arrived in a city that didn't want her, to an apartment and a business she had to fight to get back, and to a heritage stripped of all its stories. The walls and floors were bare; all the material proof that she had once had a life, a family and a sense of belonging had entirely vanished. The war burnt away millions of stories, most never to re-emerge, and those like my grandmother who had survived the odds had to start again, as if there had never been another life. She would rather have left Vienna, but she had nowhere else to go. So she bought new furniture from the auction house, took in her husband's cousin Stefan and his wife, Hilda, and started afresh.

Years later after my grandmother's death and then Hilda's, I went to the apartment on the Ringstrasse for my last visit. Along with my mother and brother, I went through the contents of the apartment, ordering everything for shipment or disposal. Faded great rooms full of faded furniture, a once grand interior worn by the weight of time. Amongst piles of linen in the dining room cupboard, each piece stitched with the initials of my grandmother, great-grandmother or great-aunt, were about forty small Bakelite boxes filled with delicate pieces of glass, dual images of another time etched onto them. I had never seen these ones before, though I was aware of another twelve boxes in my grandmother's old bedroom, neatly stored inside the viewer one could look at them through. Along with just a few other remnants of my family's prewar heritage, some letters and other documents, I boxed them up, ready for the journey north. These items had been taken from the apartment in 1940 by a family friend who had kept them safe until my grandmother's return in 1946.

The letters fit neatly into an A4 filing box, the slides take up a small cupboard, and the linen remains packed in the same cardboard box it arrived in thirty-five years ago. The viewer sits on the floor in my office, its darkened Mahogany shell, brass binocular viewfinder and dials rooting it firmly in the past, a full 114 years ago, when my grandparents were married. When I discovered the slides and other objects, it was my own personal version of the eroding of the Quinhagak coastline. The past came to light in the silence that had entombed it. The letters and pictures rolled out of the shadows slowly, taking time to look through and digest. I have been looking for over thirty years.

The viewer is a machine, a mechanical interface between the looker and the past. The dials and levers on it respond to the grip of the looker, and in my case my grip on them was akin to the grasping of a new consciousness, the opening of a hidden door. We call the viewer a *Guckkasten*, a peep box or stereoscopic slide viewer, but it is more particularly a portal into another time lost under the rubble of pain and dissonance. There is no stroking of programmed and copywritten responses, only the pressured articulation of brass cogs and carriages made by hand 114 years ago. Each depression of a lever sets down one glass slide only to collect another, to bring a partial story to life, to let slide some more of the frozen sea edge for it to spill its held secrets.

My grandfather took all twelve hundred of these slides, developed them while on the move. Many of the hidden slides I had found in the cupboard were taken on the Russian Front in the First World War. The other half of the collection was of family, particularly around my father's first few years of life, and was framed around a certain lifestyle that betrayed little of the dissonance growing around it. These covered the period from 1905 when my grandparents were married to 1935 when my grandfather died. The war pictures are a gritty rendering of my grandad's experiences, of the tearing apart of four empires and of the consequences of war on family life. They cover just one year from September 1914 till the same month a year later. To see my grandfather serving the Austrian army as an Austrian, only for his family to have their citizenship and life torn from them just twenty four years later, has enabled me to grasp why so much had been hidden in the permafrost. Slowly but surely my fascination with the images allowed me to resurrect much of what had been lost, and to rebuild a story which my father and grandmother had never been able to tell. Like the Yup'ik with their relics, I found that the physicality of the images and of the machine was able to ground and make real what had before been vaporous. What the hand can grasp becomes real, in the mass of it, in the matter and the sense of it. The material becomes material to the person who has association with it, helps to reconstruct stories and rebuild belonging.

I would wish to invite you in to look through the machine with me. It is a made object, its technology long obsolete. Glass slides have been replaced by cellulose, cellulose by digital interface, the tactile and material, by the vapour of a cloud. I have twelve hundred glass slides in various states of conservation, silver nitrate and egg albumen. I have copied another four hundred or so digitally. They are embowled in my laptop, and in my desktop, on a separate hard drive, and some are on other computers, perhaps even in a cloud. Yet if I am to invite you in, it will not be to the computer screen interface, not to the visual sense alone. I would wish for you to touch, to hear, to smell as well as to see the past as it presents itself in the present.

The box I place in front of you stands nearly 2 feet tall, much the height I was when I was first born and held in my grandmother's arms as she sat in an armchair, the box at her feet next to a Walnut chest of drawers. I would set it at eye height for you, just as I would have done for myself as a teenager, placing it on that Walnut chest. I call the box a time machine because it reveals hidden secrets. As you place your eyes to the brass and Mahogany binoculars, I help you adjust them to the width of your eyes, and the focus to a distance appropriate to your vision. I place your left hand on the brass knob, a brass needle attached to it pointing to a series of numbers from *1* to *25*. I place your right hand on a polished bronze lever. I sense in you a hesitation born of unfamiliarity, and of doubt for what lies ahead. I encourage you to press down the lever while you help the knob clockwise on its journey to number *1*.

The machine is a little worn, the clockwork cogs and gears that help the carriage slide need a little adjustment, which I have never given them, attached as I am to the particular set of movements it takes to coax life out of it. You lean forward to the viewfinder, and you gasp with surprise, turning to look at me, struggling for the words to express what you have just experienced. I have helped many people learn how to operate the viewer, have explained to them that the dual images on the glass slides construct an artificial 3D image when viewed, but direct experience is a far better teacher than words. 'I feel like I'm there,' they say as they look at the viewer with incredulity before putting their face to it again. I hang on their every word, on the description of the depth of field they perceive as they look at soldiers being fed from a mobile kitchen, at corpses on the battlefield or at my grandad proudly standing by the car he drove off to war in. I have never been able to experience the 3D effect of the slides, as I see predominately through one eye and cannot summon the other into the necessary cooperation that the viewer requires.

After you have looked through an entire box of twenty-five slides, I lift the Mahogany lid, move a small lever hidden under it and let the whole front of the machine fall forward. Though it is more subtle now, the musty smell that rises from the inner workings connects me viscerally to the story I am part of. I wind the left dial clockwise and the carriage at the bottom trundles out from under the vertical framework of the slide elevator. I pick it up from the front and gently ease it out from its tight embrace. I open the little door beneath the mechanism and take out another box, replace it with the one I've just removed. I slide the new one back into the carriage and close both doors. Hidden from our view the closing door activates a mechanism, which pulls the first slide up the elevator and places it in front of the viewfinder. My grandmother stands on the edge of a great ridge, a stick held in her hand, a mountain guide at her side and the grey shadows of the rocky escarpment disappearing behind her.

The viewer was made alongside the camera that took the images in the small workshop of the Richard Brothers in Paris. Skilled craftsmen, working by hand, cut out the brass cogs, assembled the carriage mechanism much like that of a simple clock. Brass, steel and wood were fused together through the various skill sets, and an automaton was born, an analogue machine, designed to help our hands hold the past moment and our eyes and senses experience it. While technology has leapt forward in leaps and bounds, it cannot do this, cannot invite us into such an interactive and physically engaged direct experience. It has helped me to grasp my heritage and all that remained from it in a way that makes me feel part of it.

My grandfather went to war with a chauffeur and his own car, was cared for by a batman, and appeared in many ways more like a documentary tourist of chaos than a player in it. The car was a central feature of many of the images. Amongst other paper images I have are ones of my grandmother racing cars in the 1930s and my father racing motorbikes in the 1950s, of my grandparents climbing, and my father making a speech at the annual Christmas party of the cloth factory in Vienna. I have my grandmother's yellow Star of David, the threads that once stitched it to her clothes still hanging. I have the worthless concentration camp money that she would have been cynically given in exchange for her own. I have documents of her incarceration, and pictures of family members who never returned. I have letters written back and forth with growing desperation in 1938 and '39, exploring means of escape which were never realised. Yet of all the documents and photos and objects, it is the mechanical *Guckkasten* with its brass levers and clacking solidity that binds me to my story. As my fingers pull and twist, and the carriage moves in and out to reveal new images, I feel that somehow I am a working, moving and living part of my own heritage, and have autonomy within it.

The slides mostly document people, sometimes destroyed towns or mountain vistas, but people are always central. They may take the energy of the image, command both our attention as well as that of my grandad when he took them, but they are not the only players in the images. The injured soldiers who look directly into my grandad's viewfinder are lying in an open wagon, its side woven from Willow, its carriage work crafted by the wainwright, its wheels by the wheelwright. I remember Hilary's comment about 'basket cases', the soldiers who, confined to basket chairs or carriages, were no longer good for anything else. There is another photo of a furtive couple emerging from a shed, the woman shy, the man buttoning his army uniform, the shed's waney edge planking crude and simple, hastily erected from local trees. The images show straw-braided skep beehives, farmers drawing water, soldiers sawing wood for the mobile kitchen, great iron guns standing on the edge of muddied fields, women cooking at the hearth, mending clothes and weaving rugs. Everywhere are the signs of our making, everywhere the evidence of skills then essential.

Objects carry stories and remind us of our heritage, but behind the objects' existence are the trades and skills that brought them into being. In 2003, UNESCO adopted a Convention for the Safeguarding of Intangible Cultural Heritage, including 'traditional craftsmanship'. It stated:

> Any efforts to safeguard traditional craftsmanship must focus not on preserving craft objects – no matter how beautiful, precious, rare or important they might be – but on creating conditions that will encourage artisans to continue to produce crafts of all kinds, and to transmit their skills and knowledge to others.

It occurs to me that I've spent my life trying to save my 'Intangible Cultural Heritage', and have tried to engage actively with what little remains as a means of reconstructing narratives that my family can take into the future. I see this desire within all the men and women who come through my workshop, to reconstruct through their hands an ability that was once innate, and objects that give them a sense of their own capacity. The slides root me to my heritage, as my skill roots me to the greater heritage of my human ancestry. The UK is one of only seven countries not to sign up to the UNESCO convention. In a conversation a little while ago with Julie from the Heritage Craft Association, I expressed my concerns over the potential evaporation of old skills and work practices, and the material relationship attached to them. Julie was optimistic on this front, citing the development of portfolio careers for artists, and the proliferation of courses as helping to sustain livelihoods and help positive narratives around making. The interest around making, its use pedagogically and even as experience, are all developing a new narrative for material relationship. Courses in all manner of crafts are popping up everywhere, and there is a great desire growing to reconnect to handwork.

Yet I could not help having the sense that the depth of relationship implicit in the practising of necessary craft in commercial workshops is being eroded, that the web of interconnectivity is thinner. Material provenance, apprenticeships, local relationships between makers and suppliers have all changed. I had a growing sense of a slight change in my purpose, a sense that between the pedagogy, the experiences and the trade, there is this issue around survival, the extinction, if you like, of certain trades, of a proper and honest relationship between material provenance and usage.

———

A few weeks later I am steering the van out of the shelter of the woods into the violent winds and pelting rain, tearing up the main road. There was a problem with one windscreen wiper; rather than clearing the rain to the edges of the windscreen, it pulled all the water it had pushed away right back in front of my view. I hoped that this wasn't an omen for the day, squall-soaked and blind as I travelled a couple of hundred miles around Devon and Somerset. I quietly hoped that whatever shed or workshop or office I found myself in would be wind- and water-free, so I could have a break from the intensity of the drumming monotony. I had an itinerary, a list of where I would stop, a selection of saved maps on my laptop that would hopefully enable me to find my way into the story that the day would reveal. I even had a subject for the day in my head, a sort of signpost to help my acquisitive mind grasp the material that would present itself: heritage. Though there were stops in the day that didn't directly concern my research, that word ended up informing my angle of enquiry and focus.

After nearly two hours of driving, exhausted by the weather, poor visibility and anxious drivers racing to work late, I arrived into the familiar dilapidation of straggling buildings and rebellious tarmac that were the remnants of an old air force base near Honiton in East Devon. This is a place I know, for the potholes and driving rain that seem to forever linger around it. I have shivered my way through hours standing on the cold concrete floor of the old aircraft hangers. It is the venue of a regular commercial auction, and over the last twenty years I have come here to buy much of my old machinery. Traditionally old factories, workshops, the council or the Ministry of Defence cleared all manner of old stuff through these great gusty halls.

Everything from airplanes, parachutes, archaic machinery, cider presses and council road signs would find their way here. Amongst them, house clearance items, antiques, computers and the remnants of past lives would briefly appear before finding a new home.

Now the auction acts more as a clearance venue for online retailers, the space filled with mass-produced and vaguely designed tat. These hangers have seen history vaporise into their ghostly shells, have acted as a conduit for the past to reframe itself in a temporary presence. If you come often enough, you see the same object revisit, its perceived purpose unrealised, its potential still open.

I was there to collect an old lathe. Nearly my age, worn at the edges, but solid in all its cast-iron mass. I had bought it virtually over the internet, the auction house having brought itself up to date recently. It was the first time that I hadn't stood shivering in the cold waiting for the lot number to be yelled out by the auctioneer. Nor had I seen what I was buying, confident only in the age and dependability of the machine, in its survival this far and the reputation it had had. I couldn't say that for the lamps, bedding, sinks and furniture that were already gathering for the next auction around the isolated figure of the lathe, one of the few uncollected items of the previous sale. They would soon be down the dump, more landfill accumulating around the boundaries of our consumption habits.

The guys helped me get the lathe into the van, and while I waited for them to bring the forklift over, I wandered to the edges of the hanger and found another eighteen lathes like the one I had bought pushed up against the siding. They'd all been cleared out of a local college, arrived here on an articulated flatbed and would be sold gradually over a few auctions so as not to devalue their bid prices. Seeing them there, some disassembled, other standing proud, I reflected on the colleges I had learnt in when I was younger, of the ranks of lathes and other machinery, on the availability of skills training and the heritage of manufacturing. I remembered this with great affection, for it was on the lathe that I had really learnt to trust my hands, to accept that I could make an object of use and beauty, and sense the excitement that that gave me.

I knew that those twenty lathes that had left the college for the auction were on their first trip out since they had arrived there forty or fifty years before. In all probability they were not being sold in order to be replaced with new ones; rather they were no longer needed at all, the college possibly closing or else that department finally axed. The skilled turners that had once taught from those lathes had apprenticed in factories full of many more. All those turners were gradually replaced by copy lathes, semi-automated machines replicating components from a pattern, and then by CNC (computer numerical controlled) lathes, fully automated. So with the manual lathes' demise was that of the turners and the industry they had been part of. Yet the lathes would all find homes, some like mine to be taught on, others for small businesses and others engaging people who were seeking to return to something they had a desire for, and that a working life had not allowed.

I got some lunch at a local pub whose name heralded my next venue, the Wheelwright. Once a wheelwright's shop would have stood at the entry of every village, the cart or carriage dropped off there for repair, before the owner walked his horse on to the farrier for reshoeing and then on his own to the pub in the village centre, his thirst and the necessary wait guiding him there. I learnt this story from Greg Rowland, the wheelwright I was soon to visit, but whose workshop I was struggling to locate after leaving the pub. I popped into a garden centre at the roadside and found myself in the café, a woman asking if she could help. She hadn't heard of the local wheelwright, but did I mean the tannery? A customer overhearing the conversation interjected that there was a wheelwright if I turned left out of the town centre and went just a quarter-mile or so down the road. My interest was piqued by the tannery she had mentioned, and sure enough, there was one right next door. It wouldn't have been the Oak bark tannery, the last in the UK, which I had been wanting to visit and thought was on Dartmoor? It was, but my visit to it would have to wait for another time.

I was soon standing in front of a U shape of buildings clustered around various bits of old machinery and the shells of a couple of old carts. It looked like a museum set, one where everything was purposed and used, so the patina on it was not that of neglect. I recognised the great round wooden platform lying on the grass, an iron threaded rod and turn handle rising out of it. It was here that the iron tyres would have been dropped steaming-hot onto the wooden wheel, the hissing of cooling water and the sound of pounding hammers establishing the union of the one with the other. The great solid cast-iron rollers to make the tyres stood wet in the rain over near the edge of the quadrangle. A movement at a window to my right alerted me, and on looking, I saw Greg's face smiling out of the dust-encrusted panes. He looked as though he was standing at a lathe turning, and I smiled at the serendipity of it, the old Union Graduate lathe sitting dust-free in the back of my van.

We were soon shaking hands, Greg's grip on mine making me flinch just a little. Perhaps all this writing is loosening my own grip on the physicality of things?

'Tea? Coffee?' he said, as we walked to the wooden office on the edge of the yard, apologies for a mess I couldn't see, and for the chill I couldn't feel. I was thrilled as usual to be in another working space, felt privileged to be welcomed in with such generosity, and certainly didn't see a mess or feel the cold. For me both have been a prerequisite of the creative environment.

I don't think that Greg in all his robust earthiness would have used the word 'creative' in relation to what he does. He sees himself as a tradesman, not a craftsman, as a continuum of a necessary profession that provided tools and the facility of transport for a pre-engined world. As I clutched the warm coffee, I listened to him, hearing the thoughtfulness and awareness that so often parallel the silence of so many makers' working lives. He was talking about an exhibition he was recently asked to be part of, a response to the wheel. He smiled with a degree of muted exasperation as he told me of the cartwheel he had made, and of the two other 'wheels' that two artist-makers had made. More creative responses to the idea of the wheel, the one a cut cross section of an Oak tree with a square axle attached, the other, a 3D plastic print of a full-sized wheel. He looked at me with some degree of incredulity as he told me, a maker to a maker. It strikes me that the heritage of his wheel finds its place in everything that preceded it, and that these artists' responses were some attempt to see a heritage of the future. Perhaps that sums up much of where we're at, no longer wedded to need, free to dream outside the bounds of necessity. He points out that there is no further that the wheel can really go, the minutely engineered solid wheels of the vehicle holding the land speed record having marked a perfection in wheel design.

Our conversation segues between the past and the present, between the history of the yard and the work they now do. Between his ancestry, of the De Roland they once were in a post-Norman England, and the traditions that he holds onto. I had found mention to Mike Rowland and Son Wheelwrights in the list of makers on the Heritage Crafts Association website. I knew they were one of few left in the UK, though not a critically endangered craft like many others on the site. From a time when there was a wheelwright at the entrance to every village maintaining the essential vehicles of the time, to one where there are a handful preventing the decay of the remnants of those vehicles.

At least that was some of the assumption I brought to our meeting, that his work would be mainly in restoration, in heritage preservation, in making sure that the pragmatic artisanship of old would not simply decay into the earth. 'It's not about heritage,' he said in relation to his work, 'it's about keeping it relevant.' To keep the heritage of the trade, his work could not just have relevance to the past; rather, it needed to find a contemporary relevance. He has found himself a niche building gun carriages for cadets and schools. I was puzzled.

'You mean to restore old carriages?'

'No,' he answered.

'Well, then why would schools need gun carriages, who are they firing at?' I had gone to private school, had managed to avoid being in the cadets, yet still found it hard to imagine them all firing old-style guns at one another. Greg laughed. The gun carriages that he builds cannot be fired, though in every other aspect they are functional, with wooden wheels, iron hubs and carefully crafted bodies.

'They're used for a sort of gym workout,' he said, and seeing my face added, 'You know, teams racing each other pulling the guns, working together to beat each other.'

'Oh,' I said, immediately understanding that these guns he made were a sort of rugby ball, rowing boat, a vehicle of cooperation and competition. I also remembered our yearly trips as a family to Earls Court Exhibition Centre in London to see the Royal Edinburgh Military Tattoo, where one of the displays was a race like this, two teams competing to be the first to get a gun through various obstacles. It's a niche market, which helps keep Greg in business, able to take on an apprentice who has now graduated to journeyman.

Wheelwrights of old would have sourced their materials locally much like the boatbuilders and village carpenters. Greg's dad, Mike, first started the business in the 1960s, having sought out an apprenticeship with craftsmen alive in the Victorian traditions of the trade. Work was slim, and by the '70s there were only a handful of wheelwrights still going in the UK. Occupying the same workshops that Greg does today, built around the old family farmyard, he set up a sawmill to help support his income.

With the devastation caused by Dutch Elm disease in the 1970s (an abiding memory of my childhood), Mike found himself inundated with Elm, the traditional wood for wheel hubs as well as Windsor chair seats. So a side business soon grew making pub tables, and to this day, pubs in the South West and all over the country still have the tables Mike made all those years ago to keep the business afloat and await the time when there would be a new demand for the wheelwright's trade. I have eaten and drunk off more of them than I could know.

I imagine this yard once full of cut timber drying in stick, piles of it arranged around the buildings. Greg tells me how his father can recognise an Oak butt for where it was grown, knows from its growth pattern and colour the soil on which it grew, and where in the country that was. The relationship with the tree used to be a natural part of the maker's craft. Woodland managers like my friend Mike were in close relationship with the village carpenters and wheelwrights, and the provenance and usage were closely bound to the same narrative.

As Greg walks me around the shop, we become equals, both woodworkers, he showing me his old machines with pride, and me delighted in the forest of invention I find. In the office earlier he had told me of how the community of wheelwrights remaining do not share knowledge. Rather, how they closely guard it. Loosed from the tight heritage and interdependency that would have framed trades like this, self-taught wheelwrights have adapted independent and particular methods. They are not keen to share these methods with one another. I don't find this to be the case when I am with Greg, as he generously shares information, answering all my questions and persevering when my ignorance of a technique causes me to ask again. I am not a wheelwright, and yet as woodworkers we find a mutuality of understanding.

Soon after entering the workshop and once I have fired my first childlike question – 'What is that?' – he picks up a small incomplete wheel, hub and spokes intact, but no rim. It's a lovely object, useless yet, but intriguing with its Elm hub and radiating arms. He places it onto the head of the archaic device that I had not recognised, and tightening it down, he demonstrates the hand-operated crank and sliding mechanism that allows the accurate drilling of all the spoke ends at the correct angle. A hundred or more years old, it has a slightly newer and much larger relative around the corner.

We wander outside to a lean-to at the back of the workshop, an old farm cart abraded back to raw wood by years of exposure standing beside a sparkling red-painted wagon. The wagon is a new build for a brewery, a promotional vehicle creating a historic narrative of the handmade and traditional. As I look at both vehicles, the one a workhorse of the past, the other a theatre player of the future, I admire them for the maker's craft within them. They are both fashioned through the same traditions, both essentially practical vehicles bound to basic need. Yet as I look under their bodies to their hidden structure, I am in wonder at the details no one will see, at the carved ogee mouldings and decorative scroll ends, at the repetitive indents along the lengths. I ask Greg why even on an old farm wagon there is this attention to detail, and his answer that the carving and relief work take weight off the carriage fails to entirely satisfy me. I am learning that it is typical of him to see the practical in the traditions he occupies, yet he does come to acknowledge that in that detailing lies the expression of the maker, of their skills and of their identity and pride.

Moving back into the shop, where shelves of fixings rescued from a widow's garden shed decorate one wall, we navigate other machines. The carcass of an old wagon rests on its wheels up ahead. Greg's matter-of-fact description of it parallels his approach to so much of our conversation: a lack of romantic attachment, a pragmatism, while fully articulate in the complex relationships of being a maker today. Bleached green canvas shards hung off the iron-and-wood frame, its purpose a puzzle to me, with great pneumatic tyres and a more traditional carcass. 'It's a horse ambulance,' says Greg, and then clocking my puzzled expression explained that it was for taking injured horses off Aintree racecourse. More of a hearse than an ambulance, I reflected, as the trip for the horse would not have been to the hospital, but rather to the knacker's yard. We wandered further round the workshop, tracking along the base of the U, where a strange little wagon sat awaiting completion. It was a miniature chuck wagon, a racing wagon pulled by a team of miniature ponies, an American import this, the first to be built over here. With a mix of traditional wheels and frame and the modern efficiency of roller bearings, it was yet another example of the wonderful hybrid eclecticism within the shop. Another example of Greg's determination to 'keep things relevant'.

Next to it stood Greg's new (old) copy lathe, capable of making four spokes at a time in a few minutes, hungry to be loaded with more, a pile growing to its side. Behind it, the old copy lathe stood idle, ready to be replaced by an invention of the last century rather than the one before. The shop was full of equipment, the great cast-iron beasts side by side with shaving horses, routers and piles of work in progress or work discarded. The spirit of material was alive in all of it, the spirit of creativity, of time and of purpose.

Greg works with the journeyman and gets other help painting when he needs it. All the kit allowed the two of them to tackle just about anything. I thought of the wound I had seen on Greg's finger when I arrived, the stitches fresh over a gash opened by the router. Time still rules the day, the tool chosen for its expediency, whether the shrill and enraging scream of the router or the strange quiet sloth of the spoke drill. He had told me in the office of the meaning of time, the necessity of making the business pay, of keeping it relevant not only in what it made, but in how efficiently it could make them. Time pulls him to the tools, and I see his dance around them, around the great U shape of the workshop, around the history held in his family farm and heritage.

He shows me a small hub made of Elm for a Royal Warrant job some time previously. Turned on the lathe, hollowed right through the centre and punctured with the rectangular mortices for the spokes, it has more air within its form than timber. A crack runs perpendicular to the circular hole, and Greg tells me why he no longer uses Elm. He'd received a phone call after delivery of the carriage, and on going up to London to check, found that all four of the hubs were split through like this. Embarrassed and keen to clear his name, he was up there twenty-four hours later with new ones, turned from a tropical wood, Sapele, and with the hope that there would be no more problems. There weren't and he's used Sapele ever since. Elm was always the traditional wood for hubs, as by nature it is reluctant to split, but in the damp of the Devon climate, it is hard to bring timber to an even humidity, and even Elm cannot resist splitting if it is suddenly shocked by a drier environment.

So, imported wood, kiln-dried, ends up replacing the traditional woods, the craft morphing with new demands and necessities. Greg still uses Ash for the spokes of the wheels he makes, but the rims and hubs are now in Sapele. The Ash is local, but I wonder when he will no longer be able to use it, when the dieback that is now decimating the native population will reduce the quality and then availability of it. The wheels Greg makes now are still bound by the same traditions of making as they have been for hundreds of years. But beyond the main change of mechanisation, the narrative within the craft has been turned on its head. The carts are no longer vehicles of necessity, the timber used escapes definitions of locality and arrives from the corners of the planet, and the workshop morphs its identity between many centuries, the tools of the trade adapted to speed and efficiency.

———

Leaving Greg's yard and its rich historical heritage, I headed north onto the plains of Somerset. As I drove I was remembering the conversations I had had with Julie all those months before that had led me to this moment. I hadn't known about the red list of endangered crafts, had barely considered what that would mean. I have seen so much from the perspective of the objects, of the furniture, tools and baskets, I had been less focused on the very specific trades that no longer had a marketplace.

Popping in to see her in a promotional yurt at a big craft fair, I noticed a tall man standing opposite the entrance, surrounded by circular wooden forms, wire half woven, pliers and other tools on the bench in front of him. As he chatted to visitors, I remembered Julie mentioning him to me when we'd first talked, that he had turned an extinct craft into an endangered one by becoming the only sieve and riddle maker in the UK. Steve Overthrow and I eventually shook hands, and after I'd stated my interest, we agreed to meet up in the future.

He was slightly apologetic that his workshop was a shed in the garden, and that he still had a full-time job. I was fascinated and resolved to see him when I could. The idea that an individual could be drawn to revive a trade and have the commitment to train himself through the evenings and weekends while juggling family obligations and a full-time job offered an oblique angle with which to approach the subject of heritage.

So, four months later I was on my way to the edge of the Somerset Levels, the rain not quite so heavy, but the gloom of late afternoon shadowing my every move. The new build house where Steve lives with his wife and two young children sits on the edge of the road entering the village. I'd approached down small side roads, avoiding the great arteries that weren't far away. I was glad for the calm, and glad also to have arrived, to have stopped driving. It was a little surreal after the heavy iron heritage and clutter of Greg's workshop, of the permeation of history and labour. Arriving at Steve's door, I saw no evidence of his craft outside, nor inside once he'd opened the door for me. The large downstairs space was split in two by a long, low fence, the adult sanctity and child creativity separated definitively; the children's world punctuated by toys and bubbling energy. The television burbled in the background, the bright glare of children's TV integrated in the whole. No one would have known that the last remnant of a heritage industry was burgeoning back to life somewhere within the property, that somewhere amongst the kids' toys and hubbub of family life there was a sanctuary of calm industry.

With a mug of tea warming my hands, we began to talk about Steve's life and about the journey he had been on to this point, how he had come to do what he was now doing. Laura sat between us and the children, her attention flexing back and forth effortlessly, interjecting just occasionally when she felt that Steve might have missed something. I was looking for a thread to pull, a loose end that would lead me to an understanding about why Steve had chosen to complicate his life, to add to his job and family the salvaging of a craft. I wanted to know what it was within him that he was looking to satisfy, and the social significance for him of being the only maker of sieves and riddles in the country.

My mother had owned a wooden sieve for the garden, as had my granny, and it was this association of them as methods of sieving earth that I had pictured. Steve said that any grading could have been done with a sieve, be it of earth or of flour, of cockles, coal or sand. The coal miners would be given coal for their households free of charge, but it was often full of coal dust and smaller pieces that would have suffocated the fire. They would have riddled the coal they used first, would have had an Oak-framed sieve by the coal pile. His mention of riddling sand for the casting industry jolted my memory to recollect the sieves on the floors of the great foundries next to the piles of green sand. When the cooled aluminium or bronze form was tapped out of the sand-filled moulds, the sand was pushed aside into a pile to be graded before being used again. The red heat and sweat of those cavernous spaces surfaced momentarily for me in the comfortable calm of Steve's living room.

Steve's interest in riddle making had been piqued both by its near extinction and his own love of gardening. With it came the realisation that he could make an object of practical use, a tool so much more purposeful than the cheap metal ones he had used. Tools, I sensed, were a thread on which he had pulled and, in my turn, so did I. It turned out that he had always collected old gardening tools and other bits and pieces, that for him it was the tool that was a worthwhile object. He spoke with enthusiasm of its purpose and of his desire to use, repair or remake it. On his website he has a quote from Thomas Carlyle: 'Man is a Tool-using Animal. Nowhere do you find him without Tools: without Tools he is nothing, with Tools he is all.'

When I mentioned it, he smiled. I sensed the purpose that a tool gives him. Steve is still ancient man alive with the potential of his grip, with the intention he can satisfy through making tools. The thread I had pulled at was leading me to a desire within him that spoke to me of the desire I once followed and that plays at the edges of so many people I meet. To make a tool is to feel the potential of one's creativity, to feel resourceful and empowered. I was beginning to think about his workshop, wondering when we should find our way out there and what tools he would express his enthusiasm through. I stilled the pull to ask, and let the conversation roll on, unravelling slowly like a ball of wool rolling down a gentle incline.

Chatting with him about the work he used to do, I heard a word float by in the still calm of his voice, and let it go awhile until he had finished. When he had, I questioned him as to whether the Rileys he had mentioned were the old vintage cars, and sure enough, they were. Before his present job and the one before, he had worked for a man who restored them, whose love of them drew their owners to him from all over Europe. He had learnt here about time, respect and trust. The trust the owners gave, and the lack of pressure they applied to time or money, happy only that the object of their love was being cared for, that their value of it was mirrored. He sensed in this the time he wished he could have to restore or make, to the space, peace and satisfaction that it may bring.

At this point Laura spoke, reminding him of how disheartened he had become after that job finished, then working for a steel fabricator. 'It was all about time,' he said, meaning that the pressure was always there to do something quickly, economically, not with respect or true value. It drained him, and showed him what he wanted. When he had earlier mentioned 'the Rileys', I had heard more than the name, but also the dignity and self-worth that they and their restoration had given him. He had rushed over it in his brief outline of his working life. I was glad that I had spotted it, that my father's own obsession with cars had alerted me to the name. The thread I was pulling at was rolling into quite a ball, the sentiments underlying the maker growing in its mass.

His road to learning the craft had been largely by looking on the internet for any information he could find, and by buying old riddles online to see how they had been made. Despite trying, he was unable to track down the last riddle maker, Mike Turnock, who'd retired some years before, handing his business over to a man who had subsequently died. Finally, having given up his search, a series of fortuitous events led to the two of them getting in touch. Since then Mike has acted as a mentor, happy to lend support and advice, a voice on the phone or on the web that acts as the guiding hand so necessary for Steve to establish his confidence.

It was dark outside by the time we left the back door of the house and slipped around the corner to the small garden shed perched on the boundary of the property. The rain had started again, and the wind threatened as we slipped into the low doorway, the two of us stooping under the frame. The shed was just 8 by 10 feet, benches around its edges and room enough for an old chair. It was a space for one. One maker in quiet seclusion, calm with the sense of the possibility it offered. I wonder now how many other people have been in it, again aware of the privilege of being invited into another maker's workshop. It seemed sacrosanct; that it was more than the 'workshop' that Steve refers to it as, or at least more than a workshop in the purely physical sense. This one-man space felt like a workshop for the human in Steve, an opportunity for him to reacquaint himself with the essential capacity of handwork, and through it with his true self.

In an email conversation after my visit, I had referred to the idea of men and sheds, and that he was making a vocation out of that innate desire. He responded with the story of a client who had referred to his shed as a man cave, and how he had been offended because he saw such a place as somewhere where nothing much happened. He preferred the label of workshop, a place of industry, and I understood his sentiment, 'a shop where work happens'.

He also sent me a link to a documentary about the making of an Elm-seated Windsor-style chair. There was no music, no commentary, no human voice. The sounds were only of tool on wood, and of shuffling feet. The camera sought only to document the journey of one man to make a chair from and around the confines of his small garden workshop. Steve told me how he had watched it after a hard day's work, and how in the watching of it he felt replenished and inspired. 'I think this was a part of me wanting the hide-in-workshop-and-make-stuff life!' I love the idea of that, of the 'hideintheworkshopmake-stufflife' as if it were a word. The Germans probably do have a word for it, have stuck those sentiments together so that they are so long as to be virtually incomprehensible.

To have one's own space and to be purposeful. Perhaps those are desires that we can all relate to in our search for our own volition, in our desire for something beyond ourselves yet of ourselves. Our heritage is not just in our genealogy or in the objects and stories from our ancestors. It is also caught in our yearning for what was normal, in a largely unconscious desire to reunite ourselves with the itch of our fingers, and to find in them those of our ancestors. Making in community, or in isolation, has been essential to humanity in practical terms. It is no longer so essential in those terms, as machines have taken on much of the production, but it is still vital to the human spirit, essential to our roots, to what binds us to this existence.

Steve gets into the workshop after the kids are in bed, finds calm amidst the ordered clutter of tools, components, and half-completed work. Sieves hang from the low roof space, are piled in a corner, and some sit in boxes ready to be dispatched. The steam-bent Ash or Beech frames sit sprung half open on the benchtop, waiting to be nailed together and stitched with wire. They are of different sizes, some 20 inches across, and others small enough to fit over a mixing bowl. Each needs a form, a tool on which to bend its hot steaming frame, more tools to set the spacing for the wires and determine the holes. On fine sieves, the wires are a millimetre or so apart, on others a centimetre or more. Some have a pre-made mesh held in by a second steam-bent band, while many are entirely hand stitched with wire of different grades. Some are wired in copper, objects just as much about beauty as function.

Rolls of wire sat under the benches, long rolls of bespoke mesh lying amongst them. We spoke of the mills still rolling out that mesh, happy to provide the relatively limited quantities that Steve needs. We spoke as one maker to another, each with the wonder of the making process woven into his own fabric. Yet however tight that weave is, it is difficult to encapsulate it in words, easier just to observe the relationships between the objects, tools and smells in the working environment. As Steve picked up a small bunch of innocuous red-handled pressed steel tools, the look on his face encouraged me to look closer, to see beyond the mass-produced stamped-out steel plate. He alters each of these by hand, refiles and adapts them to his specific use, the capacity as toolmaker emboldening the spirit of the maker, the maker's products the richer for it. The individual act of the hands to make an item, to cut, carve, drill or shape, acts as a message back to self around one's capacity. If that sequence makes a tool which aids the completion and ease of many more processes than without it, the sense and potential of that capacity grow. All of Steve's tools simply aid the totally handmade aspect of what he does, ensure his accuracy, safeguard consistency and allow more calm for the making sequences.

The body of the riddles and sieves are all wood, all made in Beech or Ash, which has replaced the Oak more commonly used in the past. The timber will not bend by force alone, will not succumb to the will of the maker without the aid of a third player. Heat helps the fibres of wood stretch, allows the length of it to grow as Steve stretches it into a circle, the inside face compressing, the outer circumference elongating. Cut to just a few millimetres thick, the grain of the trees' growth paralleling the length, any inconsistency can cause the bending wood to split, and any fragility of grain structure to crack. The word 'stress' is used for us and our ability to handle ourselves in a busy world, as it is used for materials that are subjected to it. They become stressed, as do we. A massage or rest can mitigate our stress, and a good hot steam can mitigate that of the wood strips that Steve pulls around his forms. Yet the margins are delicate, the wood's structure and thickness, the amount of time steaming and the desired radius all mutually dependent. Steve's first attempt at bending took twenty goes before he was successful. Changing the timber, its source, the dimensions and time helped him find what worked for him, the limitations of his relationship with the task. With no heritage of experience to work with, he had to define a new rule book and internalise sequences of movements and decisions that formed new ground rules for his endeavours.

When I left his house some time later, back through the living room and family life, I knew that Steve was reclaiming a deeper heritage for himself and his family, a deep connection to himself and to what is material for him. It was as if in creating the riddles and sieves, he is creating a bespoke mesh through which he is riddling himself, retaining what is necessary and losing the peripheral. He reclaims a lost heritage and in doing so acts as a model for all of us who yearn for connection with the physical aspects of our potential.

Chapter Nine

Yearning

To be wilful, to be of the wild, ragged growth on the margins of the main path – that is not the way our civilised society wishes us to be, with knives and forks and no hands, with our desire tucked tight up against the table's edge in case it should burst out. To desire, or be desirous of, has at its heart a slight sense of danger, as if it were clandestine, furtive, not entirely allowed. We keep our desires to ourselves, allow them out only in the shadows, and are sometimes shamed by them. To desire, to yearn, to reach for a sense beyond which we inhabit is a human inclination. As our hands have become strapped to keypads, as the expectations on us are measured in terms of the money we might earn, then what is the place for our yearning?

The rough texture of a tree, the bark showing itself on the edge of a waney plank, the smell of wood shavings on a workshop floor; these are for me symbols of my yearning, of the place where some of my purpose is found. As Lin Lovekin so clearly puts it: 'People who work with natural materials have some sort of love affair going on with what they work with. It's not just skills, or wanting to make, or being good with their hands. It's a sensual thing.'

Material

Three years ago, excited to be making some new pieces from my beautiful dried hardwood planks, I decided to build another space next to my workshop. It would be a multifunctional space that could serve as assembly shop, store and gallery. The main shop was increasingly being used for teaching, and I was forever having to clear my work out of the way. Over the next two months my friend Dave worked on it with my son Oscar, who was keen to learn practical skills, having decided not to go to university. After those two months, with the building framed and the roof going on, I got involved, and eventually Oscar and I finished it together. At times Dave's son Max came to work alongside us. The building survived the fire that destroyed the main workshop next to it, and was a teaching venue before finding itself totally possessed by me and my tools. Max sadly died in a tragic accident last year, and the building will forever be in memory to him and to the yearning of the young.

This morning I walked through the wet air and tree cover from the bottom car park up to the café, where I write in the murky morning light. I'd driven past Dave's old house moments before, and I remembered our time building the studio together. Two fathers working with their sons – two young men, the one in his early twenties, the other only fifteen – sponsoring them into their manhood. A lovely building going up where two Beech trees had been felled, local Larch posts and beams rising out of the old mulched ground. It was a beautiful time.

These thoughts led me to other young men I have met over the years who have willed something more for

214

themselves than the monotony of what they were taught at school. While some do manage to thrive within the uniformity of mainstream education, there are many who feel lost, who shut down, disappear or struggle to feel purpose. The industrialisation of the educational system grew out of industrialised thinking and practice. Art, drama and craft opportunities are cut from the curriculum, and the hands have ever-fewer places to practise what is essential to them.

Gabriel has been working with me for the last three months. Whenever I watch him at the end of the shop, his hand tools laid around him, his focus with the movement of his saw hand through the wood, I see in his silence so much excitement. The atoms of his matter are flying around, colliding in delighted energy, excited, active and alive. Having grown up in rich enclaves in South Africa, with a privileged education and countless opportunities but little access to the wild, he is now finding the wild in himself.

That spirit which we call creativity is only part of our wild inclination. Would we really wish to stop it, to replace it with the 'civilised', the sanitised, the controlled? Gabriel climbs most days, his hand's grasp essential to his life force, pulling him into his own volition, and to establishing his limitations. At his bench it is much the same, the potential within the saw and the chisel tempered by the limitations they provide, by their reflection back to him of the limits of himself.

I have worked with many young men over the years, and though each relationship has been unique, there is one that I failed entirely. Graham had come to me with his father, looking for a making opportunity. The father's concern for his son's well-being screamed out in the bleak concrete-block interior of my old workshop. The one time that this young man had ever felt requited was when he had done art, or made things. Deprived of the structure of school, and without any guidance for his creativity he had remained locked in his room.

They had arrived at my door, full of hope. In those days, we were a busy shop despite only two of us on the tools, so a pair of eager young hands was welcome both for the help to us, and my belief that it would support him. I remembered the desire I had felt at his age, the incarceration of potentiality and the sense that my hand might offer me an opportunity to find the key.

Graham joined us in the shop and for a year Kelvin and I supported him in his will for connection, and he began to thrive. Yet he struggled with the demands of the workshop, and with the time it took to acquire new skills. His body conspired against him, his hands provoking the tannins in the wood to stain it and them blue. He struggled to free himself from his feelings, and despite his resolve he never could find the connection he yearned for. He kept seeing the limits of himself, and couldn't find a way to move the posts that supported them, to challenge what he thought he was capable of.

We were unable to find a way to improve his low self-esteem; a commercial workshop environment and the exacting nature of the work itself weren't conducive to this young man's emotional development and artistic growth. With great sadness I had to let him go, as Kelvin and I were spending more time nurturing him than we could afford. The father came to collect him one last time and was angry on behalf of his own fears for the well-being of his son. I have forever felt that I failed that young man.

I was once a version of him, of Gabriel and of so many other young men I have met. I made the decision thirty-five years ago to leave my university education behind me without following any promised road it may have offered. I smelt out a sense of something else through the gifts my grandad had bequeathed me and through the opportunities my dad's ramshackle workshop had provided me. Playing about without much direction or motivation, I didn't realise how lucky I was.

At school, despite being uninspired by most of the education, I spent much time in the School of Mechanics, an amazingly well-equipped set of workshops staffed by time-served remnants of the old systems of workshop apprenticeships. I learnt to turn metal and wood, to carve, to coppersmith and practise basic joinery. So when I did decide to become a furniture maker, I already had a foundation of sorts. Not so for many young people I meet today, who have a yearning but no bedrock on which to ground it.

Last year I had finally plucked up the courage to get rid of a couple of bits of furniture, which I'd been nostalgically hanging onto. They'd come out of my grandmother's apartment in Vienna after my great-aunt Hilda died. I'd had them ever since, but their size and dark Walnut patina no longer suited our house, and they'd been stored in the shed for too long. I drove them to a local social enterprise that renovated and repurposed old furniture that would otherwise have found its way to the dump or to the auction.

I had rung earlier, and after finding my way around the back, past Bernard Leach's old cabin, I pulled up the van and was met by a middle-aged man and a young lad. The two were eager to see the new arrivals and delighted by the quality of the pieces, hand-cut Walnut veneers and hand-dovetailed drawers. The older man was clearly guiding the younger in the arts of restoration, and the two of them emanated a conspiratorial excitement. When I drove away I felt satisfied that I had moved the furniture on to the right place, that it would serve a much greater purpose than it had in my home.

While there, I let out that I was a furniture maker, and the young man's ears had pricked up. We talked for twenty minutes, he wanting to connect with a furniture maker, I fascinated by the desire he had and why it meant so much to him. He told me he had struggled at school, ended up spending much time at home alone in his room, slipping ever more into a state of depression. For two years or so he was unable to get outside, mired and frightened to leave.

Then an apprenticeship opportunity at the local social enterprise had opened up and, having never done anything with his hands before, he took the position and found his world beginning to open up. He joined my evening classes for a little while until I closed them to rebuild the workshop. He's been champing at the bit this last year to sign up for the long-delayed new classes. He has kept true to his yearning, to his wild inclination. However much he finds his way blocked, however many doors close, he keeps true to his desire for something more.

Yearning

———

With these musings on my mind I was left wondering how a desire for more connection with one's hands can be encouraged to take root. If the education system does not supply the answer, then where could I look to experience an environment where the earth was fertile enough for the roots of yearning and learning to bond?

A few months later I was on the road north, a sheaf of printed papers on the van seat next to me, a bed in the back, a box of tools, and my notebooks and laptop. Discussions held months before, emails and brief meetings had inspired my four-hour journey across the South West of England. I had clutched at the sheaves after turning off the motorway, printed instructions that guided me down country roads, through right turns and left, leading to ever-smaller lanes, passing through woods and wild fringes, and finally arriving at a gate and a carved wooden sign. I was on the edge of the Cotswolds, Cuda's Wold, and wondered what I would uncover within the world of the ancient goddess of healing.

I drove into a great field, home only to grass and a few tents, a handsome stone building and a solar-collecting barn. Families playing or cooking watched me pass down the dirt track that crossed the shorn grass and up a ridge to the invisible edge of another space. Order slowly dissolved into chaos, tranquillity to a growing chorus of colour, as a new landscape of hundreds of tents and vehicles filled the horizon. The van tripped its way across the loose rubbled track, ever closer to the live edge of the field. Trees grew all around the margins, a woods wrapped around pasture, 20 or 30 acres of grass on which the only grazing was done by humans pursuing their endeavours.

I found a space for the van off the edge of the lane, got out and set forth in the direction of the barn and a gentle hum of sound that bordered it. A sign announced that I was in the right place, and passing through the gate from which it hung, I heard the hum grow in pitch, condensed to a single voice and the quiet muttering around it. There was a large circle of people gathered around a fire, a single man speaking from the centre, welcoming everyone. I had arrived at the craft camp Bernard Graves had been running for twenty-three years, a testament to his vision and passion, and his belief in the importance of handwork.

The circle's focal point began to shift to circles within it, the tension of the whole pulsing out into multiple smaller replicas. Each formed around the voice of a single man or woman, who upon invitation stood forward to read a list of names. Individuals sheared off from the whole to orbit their new star, and soon there were multiple suns with orbiting planets, the now solitary figure of Bernard at their centre. Having identified Rich earlier, I headed straight for him, introduced myself. I had found my own star and joined a group of nine others in its gravitational pull. Rich and Mark would be leading this group, introducing them to green woodwork, to the making of stools and whittling of wet wood.

I had volunteered to help so that I could witness the event, and perhaps find in it some answers to my questions. Rich and Mark were both woodsmen and woodworkers, facilitators at a nearby centre that worked with special needs adults and was based around the teaching principles of Rudolf Steiner.

Soon we had all said hello, explained ourselves minimally and headed out to the woods, collecting a bucket of tools on the way. There were two Steves, a Janice and Greg, Alice, Alan, Dave and Sarah. The bucket of tools that I shared the load of with Mark quickly grounded me, their poking heads and bodies calming and centring me. We crossed the margin of field and woods, of the managed land and the wild, of light and dark, and the woods' edge sealed itself behind us, as if what lay there had never existed.

Beyond the woods, beyond the now deserted fire, was the intention that had brought me here. The woods was the cry of the wild, the focus of our endeavours over the next few days. The members of my group were going into it to extract and convert material, but in doing so, what changes would they experience?

This group was one of fifteen or so that had spiralled off from that fire, into various patches on the margins and edges of the field. They were called by iron, copper, animal hide or grass, by Willow and clay or by the raw colours and aromas of the earth. Young and old, over the next few days they would escape their busy minds, engage their hands and – out of their laughter, talk and the repetitions of their fingers – emerge with a treasure trove of craft. I was here to observe this change and to learn what they would take away that they had not arrived with.

The Pyrites Summer Craft Camp was Bernard's dream, his innovation, and had been going for twenty-three years.

It grew out of his lifetime connection to the pedagogy of making, inspired in part by the ideas of philosopher Rudolf Steiner. The adults and children now scattered at the field's margins were here to find a connection to the materials that called to them, to the materials that might in their transformation say something about the nature of their maker. The malleability of clay, resistance of wood, temperature alchemy of steel and fluid tenacity of Willow would work their own magic on those that had chosen them.

Some would grow frustrated with their chosen material, would feel themselves fighting its nature as they struggled to slow down and adapt themselves to it. Fingers attempting to coerce braided straw into the structural solidity of a skep, a body acclimatising to the use of a new tool echoed the internal struggle to accept the pathway of learning a new process. For all the adults and children now gathering around their chosen craft, the next days would prove to be a journey through the very essence of themselves, and in the resistance, fluidity or stubbornness of the materials, they would face some of the same qualities they, too, possessed.

The children running excitedly around the legs of the adults, out into the wild margins or clinging to a hip and shoulder, were here as part of family units. Some as young as eight would be doing one of the crafts – copper beating, blade making, woodcarving or pottery. The younger ones would spend their time in the forest school environment or in the crèche. Amidst the camp the free spirit of these safely held kids would weave a lively and energetic calm. For now they were finding their place, making new acquaintances and gravitating to others they'd met at the last camp.

As Mark and I left the field and crossed the margins of the wild with our trug of tools, I briefly pondered the sense of separation I was experiencing. I had undertaken to help out with the greenwood course, to assist Rich and Mark, and in doing so I hoped to feel a part of this camp for the week, to earn my place here and to collect stories and observations. As I unloaded familiar tools on the forest floor, I wondered if I was going to end up limited by my comfort zone, too at home with the skills and unable to experience some of the challenges that the others here would face.

We set off with a few pruning saws and a great old American cross-cut saw. I chatted briefly with Dave, learnt that this was his first trip to the camp and that he was here with his wife and three kids. In his daily life he was an accountant; here he could leave behind that identity at the margins of this land.

Mark and Rich gathered us all and talked of what we would be doing, of the woods, the Ash tree we would fell and cut, and later split and shave, and finally assemble into stools or chairs, turned components or decorative objects. The woods, on its tumbling slope, was a mix of all sorts: Holly, Hawthorn, Sycamore and plenty of Ash. Amidst the wild growth were some more mature trees, Oaks of a decent girth, a hundred or two hundred years old, their canopies rising amongst the younger self-seeded trees. Distributed all around were multi-stemmed Hazels, unmanaged coppice of another time. They were quite old, a relic from when these woods were managed for firewood and woodland produce. Mark referred to the woodland as having been 'a way of being and living', part of the then necessary relationship between nature and human production.

Rich was pointing up to a couple of whippy Ash trees, their small crowns swaying at some height as they caught the breeze. They were close together, each of an 8-to-10-inch girth, clean and straight on their lower growth because of their race to find the light, but crowded tightly into a mixed medley of other trees. He chose one, the better for its position for felling, and we began removing a few small limbs from surrounding trees to aid its clean fall. Saws cut through small branches, excited hands pruning away.

Steve and Alice took to the two-handed felling saw. Five feet long and 8 inches or so wide at its deepest, it curved along its length, diminishing in width towards the handles. Mark told the story of it, of how it had come from the States with a forester called Lucy, an older woman long versed in the care and use of such a blade to manage trees far from roads, where its light weight and portability made it the tool of choice over the hungry and noisy chainsaw.

The value of a saw like this was equal to the maintenance of its blade, and to the skill of the saw doctor who cared for the teeth and would always have accompanied the sawyers in the days before chainsaws. Lucy developed those skills by necessity, and had passed them on to Mark along with the saw. He was slowly learning what it took to keep the blade as true as it had been when he adopted it.

A sharp saw sings in a pitch that a blunt blade can never mimic, the cut spirals of wood falling cleanly from it, conserving the energy of the sawyer. The saw doctor and sawyers were not the only essential members of an old sawing team, however. The cook was material to a team's efficiency. Cooks' wages at the time reflected their critical role in the physical and mental well-being of the men. Janice turned to me and joked about her afternoon as a volunteer cook and how it had no pay at all.

As Mark shared the saw's story, we could feel his passion, the importance he placed on this saw and on Lucy's knowledge, and how in telling it he was handing us a gift. His yearning and enthusiasm became our own.

Steve and Alice were crouching now, ready to take their first cut. Mark and Rich had already demonstrated the stance, the movement of the whole body braced on widely spaced legs. The arms did little work, acting only as a hinge between body and saw as the hips danced on the pivoting motion of the legs and swayed the body to the rhythm of the cut. Our two fledgling sawyers struggled with the blade, their shoulders and arms fighting for control, their bodies and legs fairly static.

With time, they became lighter, their whole bodies, legs and arms more united, and the rhythm of the dance slowly emerged. Mark occasionally re-demonstrated the movements, his body strong and centred, its vitality a reminder of the engine that we can be, the truly sophisticated muscle mechanism that we are. Furrowed brows relaxed, pursed lips opened into smiles, and stiff legs became fluid.

Our sawing team cut to the hinge, creating a pivot which would guide the trunk safely down between the trees. The saw's teeth sliced away the remaining wood, the tension on a rope attached to the tree keeping the saw from binding, and slowly, with a guttural rumble, the trunk hinged on its base and then fell quickly between the straggling branches. It snagged briefly, but a few strong pulls and it was down. A chorus of the wild rose up into the small clearing of light that had opened.

At our feet lay the trunk of the tree, saws dimensioning it into useable sizes, the brash neatly tidied, the reduction of tree to timber well on its way. If we let it lie, it would rot back to the earth, its mulch feeding the plants and creating habitat. As it was, the Ash came back to our woodland camp, cut into lengths suitable for the stools that many participants would make. The pile was small, a reduction to human order of the spindly tree that we had removed. As we left the woods for lunch in the barn, the mood was quiet and reverential.

Emerging out of the woodlands, onto the plains, onto the running kids and blinding light, I reflected on those woodland men of another time, on their dark habitat and the nature of their days. In stark contrast, the field was full of noise and a riot of colour, the ringing of steel, the odd angry roar of a cordless angle grinder and the gentle hammering of a mallet.

Wandering around the camp after lunch, I perused the various other workshops as if I were a shopper in a supermarket aisle, taking a look at all the potential choices I had. I wanted to get a sense of what was there so that I could go back later, when I had more time. I watched the industry of hands at work, of bodies and minds engaged with making. Some 250 people, young and old, all here to make connection to their hands, to their yearning and with one another. Their desires were shared by fifteen or so professional makers, some of the many drawn increasingly to teach and pass on their skills. Skills that are often difficult to make a living from.

In Sammy's group they were already forming clay vessels, adults and children working together and separately in quiet focus. Over the days they would make bowls, cups and platters, and two kilns with which to fire their work. Moving along, I passed the wood-burning cooking area, two fires roaring, great pots already on in anticipation of dinner in a few hours' time. On brick plinths, massive iron grates bore the huge steel vessels, ready for the cooking of the meal ahead.

The ringing of hammers followed me as I passed by the forge, wrought shapes emerging from the red gas heat, students manipulating and forming them. The soft silence of leather-work wove a gentle calm over the next two spaces, groups of men, women and children chatting within the tranquillity and repetition of stitch and cut. Shoe-shaped leather cut out and laid flat awaited its form, while the gentle tapping of mallets impressed traditional patterns into thick hide.

Further along the woods' edge, there were two shelters close to each other. On the left rang out the striking of steel on iron, its pitch just a little higher, the beaten iron thinner and harder. Blades were emerging on anvils and in the red heat of small charcoal-fired forges; the essence of the blades' significance etched into the studious faces of the participants. To the right, in a state of still quiet, bows were emerging out of Hazel branches cut from coppiced growth. So-called self bows, they were made directly from the raw material and not the laminated sophistication of modern manufacturing. A few words spoken with Phil, the instructor, resonated deeply with me, and I left excited by the promise of more conversation later.

Hearing a voice call me, I turned to see my friend Debbie crouched low in a large tent behind me, her ground covered in straw, her hands guided by a grey-haired man. Only women in this tent, other than the instructor, Paul, of tousled hair and slightly mischievous smile. I stepped forward into a space quiet with strenuous endeavour, where delicate hands twisted together multiple straw strands and wound them into coils around the base of emerging basket forms. They were all making skeps, traditional basket containers that were once used as beehives throughout Europe. Seeing them there reminded me of a couple of photos from the collection that my grandad Walter took in Poland in 1915. Sitting in a simple rush-roofed shelter are two rows of these skeps, ten in all, mud-daubed, their rustic sensibility quite beautiful in the old black-and-white shot.

As I walked on down the field away from the main grouping of workshops and towards the very far end, I reflected on what I had seen. Most striking was the quiet and intense application. The age of the participants ranged between eight and eighty, yet I saw no evidence of boredom or overt frustration, only a gentle commitment to process, a steady repetition of movements and over the days to come, a growing pride in what it was they were making.

I followed the sound of ringing to the bottom of the field, where a group of three remaining workshop tents fringed the edge of the lower wood. The first was filled with young kids between the ages of eight and thirteen; their instructor, Melissa, stood quietly observing them, interjecting only when necessary. A far cry from a DT (design and technology) workshop in a school, here red-hot steel was plucked from furnaces by small hands, held and bashed by others. Assisting Melissa was a fifteen-year-old boy, a seasoned practitioner, an artisan in the making. Trust was in the air, and with it the young moved forcefully into an attentive relationship with their tasks, honouring themselves as they were honoured by Melissa.

A conversation I had a few days later back in Devon with a handful of local craftspeople further coalesced my thoughts around this. The artisans are limited by schools regarding the tools they can take in when they run workshops, knives particularly raising fear and evoking grim associations. The kids end up whittling inorganic wood with sandpaper, any imaginative construction possibilities reduced to premade kits of parts. This precaution is designed to obviate risk – and in so doing takes away the heat and power of character formation and kills desire. Risk is the definitive element in a maker's moment-by-moment relationship with their craft. Risk is also essential to our character development because a life spent within a self-imposed comfort zone never stretches our potential, never draws upon the untapped capacity within.

I meandered over to the next tent, the hammer blows from it reverberating with the greater bass of the material being struck. In the forge it was metal on metal on metal, and now here amongst the chips and shavings it was wood

on wood, on metal on wood. The blade of the carving chisels that cut cleanly away from delicate hands could have
been forged next door, quenched and hardened, handled
in a piece of Ash or Box from the woods. I watched the
faces of all the participants as they tapped and pushed,
struck firmly or light. I watched as hands felt their blades
into the crevices already revealed, slowly peeled away the
layers to allow their ideas to emerge through the surface
of the timber. Faces etched with concentration, minds lost
in the action of the hands.

Axel stood by as quietly attentive as Melissa, his years
of making stitched into the fabric of him, hands creased
with use, a calmness of character honed by a lifetime of
patient application. He pointed with great pride at the
work unfolding on the timber boards in front of everyone,
a pride not his own but for them and their efforts. He
seemed to me like a father watching his children learn
through their labours, through their frustrations and their
joy. Chestnut trees became carved signs to ward adults
from teenage doors. Alder revealed the green man previously lurking within it, and the resistance of Oak didn't
withstand the energy of older hands as they pushed into
its hard, crisp grain.

Time was soon pulling me back up through the field,
past the smoking fires and into the shaded underworld
of the woods. The scene had changed in my absence,
the atmosphere of reverence for the fallen tree was now
replaced by the bustle of industry. All hands were on deck,
dragging bits here and there, erecting shelters, assembling
the pole lathes and positioning the shaving horses. The

woodland was becoming workshop, mechanisms and machines occupying it.

I went back to the van to retrieve my contributions, a shaving horse and trug full of tools, a shopping list suggested by Bernard to supplement what Mark and Rich had brought. I helped with the lathes, reflecting on their simplicity, on the lack of iron and motor, on the technology of a sprung branch, which together with the turner's foot would act as engine. I remembered how in my brief apprenticeship I had learnt on this lathe's modern descendant, an iron Victorian treadle lathe. Whether it is motor- or foot-driven, the lathe is a device that takes some risk out of the making processes and brings a level of conformity and predictability to the end product. The natural character of wood being turned on a lathe is lost to the efficiency of the design. The object becomes ever more a product of humans, and less of the natural world.

On the earth floor near the lathes stood six shaving horses, simple bench seats with a foot-operated clamping mechanism that allowed for the firm grasp of split wood while it is shaped by a drawknife. Over the next few days some of the students would be drawn to the lathe, and some to the shaving horse. Some of the stools would have turned legs, and others more earthy and natural-shaped knifed legs. Each of the stools would end up with a character unique as its maker, some veering more towards a language of the workshop, and others to the hedgerow and wild wood of the materials' provenance.

The end of the working day had arrived, and after tidying up we all hurried to the dinner queue, having learnt at lunchtime that the kids were all super quick off the mark, and not only with their first helping. A family I sat with had been coming for seventeen years, the two girls now grown women, the eldest with her boyfriend. They knew many of the others here, as for them it was a chance to reconnect with one another as well as with their hands. Each year they tried new skills or honed ones already tried, missed teachers who no longer came and embraced the opportunities new ones brought.

This camp was a break in their year from city life and employment; it was an opportunity to retain a rhythm of connection to themselves and the community of others of like mind. These conversations continued through the days, outside the working time and particularly to those drawn to the early-morning vigil of the wood-fired hearth. Here I met Heather and Ben, regulars to the husbanding of the fire, and as we watched the dancing flames over the coming days, we chatted freely.

The next morning back in the woods, one of the Steves was fighting his length of Ash. Knots at the end of it resisted his attempts to split it. The axe he held through the centre of the round length bounced off the twisted grain as he struck it with a mallet. His frustration was growing. As a music producer of famous bands, he was used to resistance, I imagine, to compromise and negotiation. We cut a piece off the end to get rid of the knot, and soon he was flowing again, the quartered wood splitting away from the axe with ease.

Earlier Rich and Mark and I all looked on as everyone picked up a mallet and an axe or froe, and watched their delight or exasperation when the wood split cleanly or not at all. As we wandered around, we witnessed each cut length of tree slowly reduced to quartered lengths of split material. The rings marking the years of growth were now exposed to the air, visible along the two inner faces of each piece, the bark still clinging to the outside.

The bark and inner cambium layer of a tree protect and feed it, are the living edge of the tree's growth. All the years of past growth are just that, the past, as each new year imprints its character onto the growing tree. Into each ring, each thin, dark line of condensed winter's growth, and each broader, paler line of summer is a geographical and meteorological image of that year. It is a profound moment for a woodworker, to be the instrument of this transformation, the arbiter for what that once living form will become. Piles of split material accumulated around everyone's feet, and with it the atmosphere relaxed, the human chatter thick around the mallets' crack.

Those split lengths had found themselves in a liminal place. They were no longer tree, and neither were they yet transformed into an object which had absorbed a narrative, giving them value and worth. They were in limbo, between worlds. I have watched many students look at these sticks with a sense of incredulity, holding back the inclination they have to devalue them, trusting only the confidence of the facilitators that they will become something more than just sticks. What unfolded from this point was always miraculous: as they become the ones to shape the new narrative for the wood at their feet, they become the agent of the wood's future, and transform a little of themselves in the process.

The Schumacher College is relatively close to where I live, its students coming from all over the world to look at how they can contribute to the well-being of the planet and every living thing on it. They have often had conventional jobs, and are bright and enquiring in their search for a different path. I have the privilege of spending a few days with those students on one of the master's courses every year.

They come to the workshop, work together as a team, learning green wood skills, and make a bench. The benches of the last few years are scattered around the college, two large ones standing outside the main entrance, their natural forms silhouetted on the bright white wall. I see the students occasionally over the course of the year, bump into them at the college or on the street. Many of them talk about those benches and how important it was for them to find another way to be in their bodies, to realise that another level of thinking would emerge through their hands if they let it. The benches stand as testimony to the students' experiences, reminding them of that broader sense of their own capacity whenever they walk by.

The students in the woods are sitting on the shaving horses now, holding their split lengths in the jaws that clamp tightly shut with pressure from their legs. There is no screw, no mechanical vice, only them pulling the drawknife along the grain of the wood, the force of their actions relative to the clamping pressure applied. The woods are quiet; Mark, Rich and I idle, and time quickly disappears through the day. Sharp knives slice through the wet wood, resistance fades into ease and the rough legs find their human expression.

When I first met Bernard some months before, he had been surrounded by deerskins, which he was stretching. Hanging in wet translucence on wooden frames, they in turn framed our first meeting. I'd helped him carry the hides from the shading front of his house to the sunny rear, and after we laid them out in the back garden we chatted in their midst. Those skins were for an event like this one, and as I wandered the field later, I saw ones like them used for making drums.

Round frames made from fresh steamed wood, bent around pegs, and the skins stretched wet onto them, to be painted later. The finished drum in all its bright colour, and the sounds it now played, no longer bore any resemblance to those stretched skins. Yet their transformation had wrought upon the drum makers a similar metamorphosis. The sinews and musculature of a human when making reveals part of the soul. The smiles and laughter around the crouched figures, the soft pride cloaking them as they each described their pieces to me. The colours they had chosen, the designs and patterns, all held a personal significance. The makers' memories of this time would never leave them, and the drum hung on a wall or propped awaiting use in a corner would forever speak of the experiences of the week in which it was made.

I soon found myself pulled back to the bow-making workshop and to the instructor, Phil, an ex-army man whose quiet wisdom and calm energy welcomed me in.

Bows, once used under royal decree, are now just for sport or for moments like this. Their making, each stroke of the hand drawing a tool over the raw wood, draws the students' attention towards themselves. The tillering of the bow parallels our gentle path towards self-knowledge, the tillering of ourselves on our journey through life. Tillering is the process in bow making of conditioning the body of the bow and seeing where it calls for more wood to be removed. The repetitive flexing of the bow stretches the fibres and reveals how efficiently it is bending. The tiller on a boat guides sailors to their destination, and as I watched participants from twelve to seventy pulling rhythmically on their self bows, I couldn't help but imagine that this process was somehow guiding them too.

Phil held the space in quiet confidence. There was often surprise etched into students' faces or conveyed in language; surprise that they had been able to make the bow, the basket, the leather bag or shoes. Somehow the sum of their repetitive labour had caught them unawares – there in front of them was a finished product they had never imagined would emerge.

Back in the woods a quiet energy was growing. The other Steve was on the pole lathe, the end of his Ash leg bobbing around as if it were the head of a toy bulldog in the back of a 1970s Ford Cortina. His rhythm and breathing were steady, and slowly the bulldog's head calmed down, the Ash stick transformed to a chair leg and the slight tension on Steve's face morphed into a smile. The answers to the questions around why we make are answered in the observations of moments like this. Heady debates in academic books around the meaning of making are fascinating, yet it is only in the practising or witnessing of it that meaning is found outside the mind.

Not long before, I'd had a conversation with a lady at the camp about skill and her struggle with it, the feeling that she would never get it. Yet as we spoke, and I learnt about her academic role and the books by which her enquiry was led, I was reminded of the space between thought and endeavour. This is the point which all my students meet when their thinking around what they are about to do separates them from the doing of it. The bridge must be crossed through the engagement of the body and not the mind. Steve wasn't thinking about his stick bobbing maniacally around as he turned it; he simply stayed with the task, trusting his body's own knowing.

The other split sticks were also transforming. Chair seats upside down punctured with holes, the fettled legs growing out of them, as if reimagined branches out of a tree trunk. Dave was working out the proportions of a drum stand for one of his children, alive with the imagining of it in use, puzzling out the angles and stature of it. Sarah turned a rolling pin on the pole lathe, a lemon squeezer next on her list of domestic objects she wanted to make. Behind me the repetitive tapping as the first Steve worked the chisel across his chair seat, imagining the shape and form he would need to make it comfortable.

As I walked back to the edge of the field, I was reminded of the small shelter squeezed almost invisibly into the hedgerow between field and woods. I'd passed it every day, and had wondered who had camped in the between, not on the field and not in the woods. It felt like a hide, of the wild yet not. One morning late in my stay I discovered that it was Ben's, my companion of the early-morning fire vigil, a man of camouflage green and a hint of mystery. An ex-Marine in his sixties, he talked openly about PTSD (post-traumatic stress disorder) and getting used to men's hugs after five years of coming to this camp. In all that he didn't say, I sensed what this means to him – helping out here, being involved, witnessing everyone.

It was now seven in the morning and the camp was slowly waking up, adults sleepily following their young kids out onto the field in their dressing gowns, the kids already full of play. Amongst the children there was a girl of ten or so with Down's syndrome whose presence I had quietly been aware of all week. She was free to be herself, to engage with the others or be in her own world, free of any obvious parental constraints or concern. I reflected on how wonderful such a yearly experience would have been for my daughter, Misha, and I wished that we had come with her all her life, so that she could have taken part in the communality, safety and learning of this unjudgemental environment

As I realised that I was now on my last afternoon, due to leave the next morning on the penultimate day of the course, I felt the joy of having witnessed everyone's labours, the joy of watching material transformed to object while the process transformed the maker. However, I felt that I had missed something, and needed to experience a little of what everyone else had before I left, to participate a little and not only observe and help, to satisfy a little of my own yearning.

Yearning

When I got back to our workshop in the woods, I found Rich and asked him to show me how to make a small stitched container from Elm bark. He had a couple of them in one of the tool boxes, the one protecting a large chisel, the other acting as a holder for various items. They were stiff to the touch, felt tough as a woody leather, and were so robust it was difficult to imagine how they were made.

An hour or so later I was pulling a length of Elm bark out of the water bath we placed it in before lunch. It was soft and pliable, like the skin of a hoary elephant, belying how intractable and stiff it had felt before its soak. Rich gave me a quick explanation of what to do, and soon my afternoon was disappearing into the cut and stitch, as cut bark carefully scored and bent, punctured with repetitive holes, slowly regained its memory as a cylinder. The space within the bark, once of the woody tree, was now filled with space and potential, awaiting whatever use I will give it, and the associations that it will hold for me.

I had used only my knife, had made the thread from the bark itself, and I can't now help but wonder that something so empowering could also be so simple. I am used to the hardness of timber, the need for a family of honed tools and skills, of much time and a whole series of actions leading up to and following the making of a piece. The extraction and processing of the tree, the drying of the wood, the dimensioning and flattening of it before perceiving what it can be used for.

I am with Bernard's differentiation between 'handwork' and 'craft', the one being of soft materials easily worked and transformed, the other needing more sophistication of facilities and skills and time. All the skills learnt within the field and woods over those days had fitted more within the narrative of handwork, even the green woodwork we had been doing. The actions did not demand too much knowledge, did not lay down skill as a hard impediment to experience.

When I loaded the van the next morning, my tools packed into trugs and boxes, the bark wallet I had made protecting a favourite tool, it stood as a talisman for my own learning and as a reminder that ease of purpose is as valuable as complexity. After saying my goodbyes and driving back down the bumpy track, and out of the gate onto the lane, I felt excited about the possibility of returning with my wife and daughter. I was confident that Misha would find enough ease in some of these tasks, and enough enthusiasm in the communal goodwill that she could be equal with others for the week, feel fully included. I was also a little sad, for if that young man who apprenticed with me all those years before could have done this camp back then, when it first started and he was first searching, then perhaps his yearning could have been better supported.

I also wished that my two woodworking sons could have been there with me. My youngest son, Oscar, builds wooden cabins, vehicles and structures. My eldest son, Felix, is a spoon carver and facilitator, has a sculptural disposition which has led him to what he does, and has fostered a relationship with his craft through a desire to connect to his hands and himself. We had the opportunity just a little while ago to hang out in my studio together. He wanted to build a small workbench, and I wanted an excuse to spend time with him. He has taken his desire to make, and made a life with it, ingrained the actions of axe and knife into his emerging form. He speaks about the resources he is accumulating through his work, the materials that are emerging as he dedicates himself to a crafting life. He sees that he has been led here through necessity, through the window of his passion to find the internal and external resources that he will need to create stability in his life.

He is eloquent, and I know the voice well, that of the maker who in the making is piecing together their thinking. As we tap away together in my studio, the walls full of tools, big benches and a machine shop next door, he speaks of the need to establish these resources without reliance on the physicality of workshop, machines and hundreds of tools. He recognises the commitment that is endorsed through their accumulation, but he also yearns to be the itinerant maker, a few tools in a bag, materials harvested from the hedgerow. He wants his resources to be within him, and not caught in the trappings of a trade, of an identity. His identity is held within him, not within the constructs of what he does, yet it is by means of those constructs that it has developed. I am in awe of him and his clarity and the path he has taken after an unsatisfying education, to build a strong sense of his own empowerment through his hands. I am in awe of both my sons for this, and all the other young men and women who keep what they yearn for alive, and refuse to compromise it.

Chapter Ten

Full Circle

I wonder if the ripples in a veneer on the surface of a table can speak of the tree that they were cut from. If a tree could speak, what would it say? Are we so different from a tree, blood and sinews or sap and cellulose – matter transformed to life, and then to death and decomposition? Death hastened by an interaction with time and air, lives lived through action and perceived inaction, words sanctifying the human and separating all else. I am not a tree, nor could I be, too impatient to stand for so long in one place and allow the force of the wild earth to run through me. I wonder of the stories, of magpies that bury their dead, albatrosses mating for life and mourning the passing of the first to die. Human language imposes itself on the natural world, as do my words here, a thirst for meaning, for what it may mean to live a shared life.

I have allowed my words to form this book, and allowed them to form me as well. The materials from which I have crafted furniture, by which humans have established their dominion here, are those through which my words have grown. We craft objects with material, books with material and we clothe ourselves with material. I have found materials on my journey through the writing of this book, which have helped me understand my place. The writing process is a craft that has a direct association with our well-being, with our understanding and belonging.

Some years ago I signed up for a master's degree in Creative Writing for Therapeutic Purposes (CWTP), an attempt on my part to explore my belief in the value of the writing process as much as the value of its product. The table crafted by the furniture maker is forever a symbol of a relationship with the material they made it from, and the journey to its completion. A book, it strikes me, is much the same: a product for the enjoyment of others which represents the processes of the writer, and their distillation of the materials they draw on. The course was run by Claire Williamson, and it was to her house I found myself journeying some time ago, having not seen her for some years.

I had read her poetry, knew a little of her story, of what had pushed her to reach into her creative depths and express herself in language. I loved what I had read, her candour and the pacing of it through the constructs of metaphor. It created a fine thread of personal connection that promised something for the future. So I reached out to give that thread a little tug.

She heard me, and the thread resonated back, strengthened just a little each time one of us plucked it. I found her house on a balmy summer day, and stepping out of the car alongside a garage entangled by weeds, I allowed time to stand still, expectations to fade. A low metal gate, stiff drop latch amenable to my touch, stone steps leading me to the front door. Catching a movement to my right, the drift of words, and excited action of a wagging tail, I veered towards it and found two men sitting in the sun on the stone paving. I was pointed back to the front door, and soon Claire was standing there, as ever I had remembered her. Our warm embrace spoke of a new relationship; the threads had rewoven into a fine and strong rope. A young dog fought for my attention, and successfully commandeered a little of it.

Time spent around the kitchen table as Claire prepared a salad helped to braid a rope bridge over the three-year gap that we could cross easily. We spoke with ease of things that could not be spoken of before: of our children, of the delicate balance of our own paths with those reliant on us, of our creative edges and the place we found for them. Claire and her partner, Andrew, had moved to this old house three years before, and at lunch I heard from Andrew's father, a retired architect, of its history, the life it had lived or the one they had reimagined for it. As we ate we looked out over the River Severn, motorway bridges spanning its width, wind turbines around Avonmouth sentinels to a distant view.

We moved back into the cool of the house, to Claire's great book-lined workroom, computers arrayed on desks. She splits her time between organising the courses, teaching, facilitating CWTP sessions and writing. Moreover she's doing a PhD around the subject of bereavement, a subject that finds form in her writing. I am here because I have questions about the craft of writing, of the material that that craft draws on and of its creation of a physical form, a body of substance, an entity, if you like, similar to the product of a potter's craft.

It had struck me that as I talk about making through the experiences of others and myself, I am giving expression to this through the act of writing. The path unfolds in front of me, and there is no corresponding map for it. I follow a hidden track and find where it leads me. So here I am in Claire's office, not sure exactly what it is that I will find. I have come to Claire's front door much as I might have gone to a timber yard or to a tree lying in a field. I am not entirely sure of the form that tree or the planks from it will take. But one thing I do know is that they are material for the journey that will unfold around them.

This book is a fluid path from an idea, along a stream bed whose variations, detours and eddies are unknown until the water that flows into it finds itself moved. So, too, is much of the work I do now, inspired as it is by the material, the planks or trees, it moves from them into a notion of what it might become. But that is not exactly what it does become; meanwhile there is a trust that whatever emerges will be the right thing to have emerged, becoming the object that could not be seen but that is right.

Back with Claire on the sofa in the great window in her office, we begin to touch on this, on the substance that the writing process creates and on the effect that the process has. She goes to the shelf and picks a book from it to make a point. It is her most recent collection of poems, and she reads me one.

I had asked a peculiar question: 'If wood is the material that a boat is constructed from, then what is the material that the human is constructed from?' It felt clumsy, as if I were labouring a comparison for the sake of proving some sort of argument. That she went to the poem, to a physical expression of her internal process, was not surprising. The poem she read to communicate her response burnt into me, and will remain for much time. The experiences it captures, and that she articulates through the imagining of horses' hooves, teacups and spilt milk, are a metaphorical construct to approach an impossible subject of loss, and of death.

She had an insight while in the forecourt of a petrol station, that she could fulfil her wish for her brother and mother to return to her only if they did so as horses. The material from which she works rises from a disturbed pool, and her reassembling of it into a poem allows the chaos of emotion to find some structure, to gain a form. She pauses after reading it and wonders that she can keep coming back to it, to the articulation she found when she wrote it, and to the sense of some peace that that brings. Words written onto paper, tapped onto a keyboard, become physical expressions of the field out of which they were drawn; they become the structure of the poem. As such they are like the boat that took much wood and craft and relationship to rise out of the ground. In her poem there is an expression of her construction, of the experiences indirectly observed, and her emotional responses to them that have helped build a layered image of her which the reader will have the privilege to share.

As we delved further into the subject she spoke of how this process of writing, and the physical construct of 'a piece', made her 'real' to herself, gave her a sense of form and allowed her to 'exist' separately to the 'service' that her childhood experiences had frozen her in. Words on a page became material for her, of texture and substance. 'We've lost touch with the body of a thing, a life, when we forget ourselves in relation to the whole.' Her writing helps her reconstruct herself in a way that she can relate to, that makes her feel real.

Earlier in the conversation she had spoken of the first time she had been able to face into the feelings surrounding her brother's death, of a catkin she had found on the ground, of its open and fragile skeletal structure. Through its vulnerable form she was able to face into the emotions she had dammed up, and through that process was able to write those feelings into a more substantial form. That fallen catkin opened a pathway through which she could meet herself more fully.

———

I clutch a great paper sheaf. I hold its mass in my hands. My words, after so many months of writing, have for the first time been made physical, dimensional through the paper they now hold to. I sit in the car for a while leafing through them, flicking the pages randomly from one to another, watching where they fell, grabbing a word, a sentence, letting it fly through me. My words have become tactile, the sense within them, that they grew from has re-physicalised, has gained the substance of the feelings that gave rise to it.

It is the sense that we search for, that resides in us and can find its way out through the articulation of our muscles and ligaments. The synaptic distillation permeates our fibres, finds them as telegraph lines along which to tap out messages. The flicking of our fingers, the movement of minute bone and tendon, is an active physical manifestation of thought; a reminder that we are a sensate body and integrated whole. This book, my writing or a made piece is an extension of me, as made through me it becomes an articulation of the essence of who I am. Words matter, have matter, become material as much as they arise from material.

A conversation with a friend of mine in a café over a cup of coffee reveals in its unfolding more layers to this relationship. Pip is a published writer of fiction, a wonderful wordsmith and imaginer, and also a cook. We had started our time together over the drifting smoke coming from the chimney of a wood-fired smoker. Inside were many cauliflower heads, their cream carapaces already morphing to grey green, giving them an appearance more akin to the human brain.

When I ask him how the action of cooking and writing compare for him, he pauses thoughtfully for a moment and then replies: 'They are the same, exactly the same.' I am surprised and ask him why. 'They are the same because in both cases you create something for the ingestion of it by another.' The writing of a book creates content to be consumed by the buyer, as the cauliflower cheese that the cauliflowers will become will also be consumed in the café where Pip works. Watching him smoke the cauliflowers, realising that he is carrying in that process a sensory picture of the final dish, allows me to understand his point.

His ideas, whether in a book or in a meal, have coalesced into an experience to be ingested by another, and at that point are no longer his. His books, inspired by his ideas and experiences or by stories told him by his Greek grandmother, are his attempt to complete the incomplete, to answer unanswered questions and bring some resolution to the past. However, once printed and lined on the shelves of bookshops or the virtual ones on computer screens, they are only what the reader will perceive them as, the experience that their reading etches into him or her.

The materials of the natural world, or those of human experience, go to form objects for consumption and use. Cauliflowers become energy for the human; wood becomes a bed for sleep; liquid clay becomes a hard, resilient vessel; experiences coalesce into a book. The subjective experiences of the cook, the furniture maker, the potter and the writer become objects with a separate life apart from that through which they have emerged. The crafting of these objects has transformed the material from which they were made so that it can now have effect on or elicit transformative potential in the eater, the sleeper, the drinker and the reader.

I think of all the makers I have spent time with over the last year: the weavers, basketmakers, potters, wheelwrights, riddle makers, writers, printers, carvers and foresters. I think of all the objects they have made, and those I have seen and handled. I can touch into my experience of these objects, of their physicality, of the language that I hear, but I can only imagine the experience that their makers had while in the process of creating them.

This book is rooted in the relationship that a maker has with their work. Those roots were nurtured long before

I started writing it, through an exchange I created with the poet Chris Waters. I had been attending workshops Chris was running, and we soon realised that we had a connection through our practice. He was a writer who made furniture on the side, and I was a furniture maker who wrote on the side.

So, with an undertaking that he would support me in my writing endeavours, he spent some months journeying over to my workshop to make a chair in Yew wood for a table he had made some years before. A beautiful chair emerged through his work and my support, and its physicality laid the foundations of our developing friendship. Those few months became years of support that Chris has given to my writing, and I am in wonder of how that beautiful chair in Yew has helped support me into creating the physicality of this book, and my own becoming.

Sitting here in my studio, the cloudless sky above me, I feel held by the soaring Beech tree that rises over the skylights. I am supported and transported, on the earth with my tools and timber, yet open to the expanse of the universe overhead.

On my workbench alongside my laptop are a couple of dozen Chestnut sticks, whittled down over the past couple of days from a cut length of a tree that has lain outside the building for the last year. The brown Oak top of my workbench lies under them incognito, the labour of mine that is in it a memory of mine alone; the purpose of it now to support my skills.

On the floor ahead of me stands a bed half complete, two great pippy Oak slabs spaced by Ash lengths, the whole held in form by simple joinery, the burry knots remembering the little branches that once grew out of the tree that the planks once were.

On the walls around me are my tools. The set of Japanese chisels that I bought in 1985 from a small shop in North London, their handles burnished by use, their fluted blades worn partially away through years of resharpening. Framing chisels and slicks, their great blades dwarfing the others, their heritage older than my own. Another set, their handles handmade from octagonally shaped Boxwood, the artistry in them reminding me of my duty to mine.

Axes and adzes from the UK, France and Austria; carving chisels from Lithuania; Japanese and French rasps; spokeshaves from Canada; knives I have made myself. The great pattern-maker's vice on my small bench, its jaws swivelled at 90 degrees, its mouth angled to hold a tapered leg, its story held in the pattern-maker's shop in the naval dockyards at Plymouth.

On the wall ahead of me, dried timber racked awaiting its fate – Cherry trees from somewhere in Devon, burry Oak from outside Exeter and Elm slabs from Plympton. To their left two camel kilims, their weft frayed by time and moths, their journey here on the back of my bike from Turkey a lifetime ago. There are two planks of Yew wood standing upright in the corner, their yellowed resinous faces etched finely with their years, the occlusions of the tree reimagined as brown, black bark shadows weaving broadly through their waxy surface. The tree was only just over a foot wide, yet I have stopped counting its annular rings and am already over one hundred. It would have seen my grandad fight at Gallipoli, and my grandma just about born. Its relatives have stood as sentinels of death, and of spirits moving through life. I wonder if it will stand in the corner forever.

On the floor in front of the bed are piles of Chestnut shavings alongside the shaving horse, their aroma filling the air, their presence a reminder of the wet wood yielding under the drawknife. I am reminded to turn my attention back to the spindles I am still to make, which will soon puncture the headboard of the bed and carry its crown.

This is the first piece of furniture that I've made for two years, the first bed that my wife and I will have had for fifteen years. Soon those beautiful slabs of Oak from a two-hundred-year-old tree growing in the wild on the windy edge of Dartmoor will hold us in their quiet embrace and whisper silently in the night.

I make the bed as I finish the book, drawing in the string on this part of my life. I reflect on Claire's comments about the finding of a part of oneself in the process of writing, and on a comment of Kathleen Jamie's that she related to me. Apparently she had said somewhere that in order to write a new book, one had to make space for a new self to grow. I like that, and find my own narratives fitting it conveniently, that what the workshop fire burnt down was actually the space for this enquiry to fill.

In the process, more space has been made to reclaim a new maker's identity, to find another version of what I searched for when I finished university all those years ago, before one path took precedence over another. If the material one draws on for a book or an object fills a part of oneself that was vacated to write it or make it, then for each making, there is a remaking of self. There is a continual process of the old substance of oneself flaking away to allow the new to grow.

Words have a mass yet can fly away on the tiniest gust of air, are solid yet vaporous, have weight yet exist only in their uttering. I hold a heavy weight in my hands, a clear glass bowl, thick and heavy, my small child hands struggling to hold it. I place my thumb in the soft recess that punctures it, my index finger in the other one and grip it as if it were my own mass. I feel in it a memory of the time I first held it as a ten-year-old, a reflection back to me of the weight I wished to have. The mass of words in this book aren't so tangible as the weight of that great, thick and heavy polished glass vessel, yet they are filled with the weight of me, with the material of my own gathering.

My mother bought this bowl when she had lived and worked in Helsinki in the 1950s, and when she told me a while ago that she had sold it, I felt sad. This bowl inspired me to make, the sheer weight of it, the bubbles trapped inside; its smooth texture and ground base gave me a sense of the weight and mass I could create for myself through making.

I have never touched another object that resonated for me so deeply. I hold it in my memory as if I had it with me all the time. It is the gravitational field towards which my making endeavours strain.

About the Author

DOLLY KARY

Nick's childhood was spent brewing beer, cooking, restoring furniture and crafting all sorts of stuff with his grandad. This has evolved into a livelihood of making, teaching, and writing, while building his family home and cabins. He has spent the last thirty-five years developing his skill set as a craftsman and designer of fine furniture for clients. He is an associate lecturer at the University of Plymouth and at the Schumacher College in Devon. He also teaches furniture making from his own workshops at The Brake, the home and creative centre he has established with his wife Dolly over the last twenty years. It is here he can practise his passion for helping others find another way of thinking through their hands. He has two sons and a daughter.

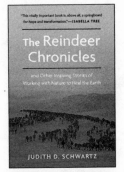